Computational Music Science

Series editors

Guerino Mazzola
Moreno Andreatta

More information about this series at http://www.springer.com/series/8349

Guerino Mazzola • Yan Pang • William Heinze
Kyriaki Gkoudina • Gian Afrisando Pujakusuma
Jacob Grunklee • Zilu Chen • Tianxue Hu
Yiqing Ma

Basic Music Technology

An Introduction

Guerino Mazzola
School of Music
University of Minnesota
Minneapolis, MN, USA

Yan Pang
School of Music
University of Minnesota
Minneapolis, MN, USA

William Heinze
School of Music
University of Minnesota
Minneapolis, MN, USA

Kyriaki Gkoudina
School of Music
University of Minnesota
Minneapolis, MN, USA

Gian Afrisando Pujakusuma
School of Music
University of Minnesota
Minneapolis, MN, USA

Jacob Grunklee
School of Music
University of Minnesota
Minneapolis, MN, USA

Zilu Chen
School of Music
University of Minnesota
Minneapolis, MN, USA

Tianxue Hu
School of Music
University of Minnesota
Minneapolis, MN, USA

Yiqing Ma
School of Music
University of Minnesota
Minneapolis, MN, USA

ISSN 1868-0305   ISSN 1868-0313 (electronic)
Computational Music Science
ISBN 978-3-030-00981-6   ISBN 978-3-030-00982-3 (eBook)
https://doi.org/10.1007/978-3-030-00982-3

Library of Congress Control Number: 2018957426

© Springer Nature Switzerland AG 2018

This work is subject to copyright. All rights are reserved by the Publisher, whether the whole or part of the material is concerned, specifically the rights of translation, reprinting, reuse of illustrations, recitation, broadcasting, reproduction on microfilms or in any other physical way, and transmission or information storage and retrieval, electronic adaptation, computer software, or by similar or dissimilar methodology now known or hereafter developed.

The use of general descriptive names, registered names, trademarks, service marks, etc. in this publication does not imply, even in the absence of a specific statement, that such names are exempt from the relevant protective laws and regulations and therefore free for general use.

The publisher, the authors and the editors are safe to assume that the advice and information in this book are believed to be true and accurate at the date of publication. Neither the publisher nor the authors or the editors give a warranty, express or implied, with respect to the material contained herein or for any errors or omissions that may have been made. The publisher remains neutral with regard to jurisdictional claims in published maps and institutional affiliations.

This Springer imprint is published by the registered company Springer Nature Switzerland AG
The registered company address is: Gewerbestrasse 11, 6330 Cham, Switzerland

Preface

The idea for this book came from one of my students, Yiqing Ma, in my course *Introduction to Music Information Technology*. She suggested that we should continue our tradition of writing textbooks for my university courses at the School of Music of the University of Minnesota, a tradition that has successfully produced the books *Cool Math for Hot Music* [42] and *All About Music* [43] for Springer's series *Computational Music Science*.

Writing this third textbook was a challenge in that our presentation of basic music technology not only comprises core concepts from acoustics and analog and digital approaches to this specific knowledge, but also includes computational aspects with their mathematical and software-oriented specifications. It was our aim to transcend a purely qualitative discussion of recent progress by a rigorous introduction to the quantitative, mathematical, and computational methods that are crucial for the understanding of what is at stake in this fascinating field of computer-aided musical creativity and data management.

This textbook is addressed to anyone who wants to learn the core methodologies of this field from the very beginning. Besides our experience with the students' learning curve during their pedagogical development in the course, we decided to produce a text that can be understood by undergraduate students of music, and not only by an audience that already knows mathematical and computational methods and facts. Our approach was therefore driven by the condition that every single sentence of this book must be stated in a style that can be appreciated by non-specialist readers.

To this end, the co-authors of the book were asked to collaboratively create a text that meets their basic qualification of interested and intelligent, but not yet specialized participants. The co-authors are undergraduate students Yiqing Ma (music/psychology), Tianxue Hu (music/mathematics), Zilu Chen (music/computer science), and Jacob Grunklee (electrical engineering/music), and graduate music students Yan Pang, Bill Heinze, Jay Afrisando, and Kakia Gkoudina.

This pool of students guarantees that the present text is accessible to the non-specialist audience. This setup of co-authors defines an approach that is more than a first presentation of our material, it is a communicative singularity that creates a bridge between a highly innovative technology and its precise and thorough comprehension. It is our strong belief that this endeavor will help fill the difficult gap between application and understanding. Figure -1.1 shows our co-author group in a happy mood after the accomplishment of our project.

This book also gives access to a number of sound examples. Here is how to find them online: The music examples in this book are available as MIDI, Sibelius, or MP3 files. They are all accessible via
www.encyclospace.org/special/BMTBOOK.
So if you are looking for the file **XX.mid**, you define the address
www.encyclospace.org/special/BMTOOK/XX.mid.

We are pleased to acknowledge the strong support for writing such a demanding treatise from Springer's computer science editor Ronan Nugent.

Minneapolis, May 2017

Guerino Mazzola, Yan Pang,
Bill Heinze, Kakia Gkoudina, Jay Afrisando,
Jacob Grunklee, Zilu Chen, Tianxue Hu, Yiqing Ma.

Fig. -1.1: From left to right, top row: Jacob Grunklee, Jay Afrisando; middle row: Tianxue Hu, Zilu Chen, Yiqing Ma; front row: Bill Heinze, Guerino Mazzola, Yan Pang, Kakia Gkoudina.

Contents

Part I Introduction

1 General Introduction 3

2 Ontology and Oniontology 5
    2.1 Ontology: Where, Why, and How 6
    2.2 Oniontology: Facts, Processes, and Gestures 6

Part II Acoustic Reality

3 Sound 11
    3.1 Acoustic Reality 11
    3.2 Sound Anatomy 12
    3.3 The Communicative Dimension of Sound 14
        3.3.1 Poiesis, Neutral Level, Esthesis 14
    3.4 Hearing with Ear and Brain 15

4 Standard Sound Synthesis 19
    4.1 Fourier Theory 19
        4.1.1 Fourier's Theorem 19
    4.2 Simple Waves, Spectra, Noise, and Envelopes 24
    4.3 Frequency Modulation 27
    4.4 Wavelets 30
    4.5 Physical Modeling 32

5 Musical Instruments 35
    5.1 Classification of Instruments 35
    5.2 Flutes 36
    5.3 Reed Instruments 37
    5.4 Brass 39

	5.5	Strings 40
	5.6	Percussion 41
	5.7	Piano 43
	5.8	Voice 44
	5.9	Electronic Instruments 45
		5.9.1 Theremin 45
		5.9.2 Trautonium 47
		5.9.3 U.P.I.C. 47
		5.9.4 Telharmonium or Dynamophone 48
		5.9.5 MUTABOR 50

6   The Euler Space 51
   6.1 Tuning 51
      6.1.1 An Introduction to Euler Space and Tuning 51
      6.1.2 Euler's Theory of Tuning 52
   6.2 Contrapuntal Symmetries 56
      6.2.1 The Third Torus 56
      6.2.2 Counterpoint 59

Part III Electromagnetic Encoding of Music: Hard- and Software

7   Analog and Digital Sound Encoding 65
   7.1 General Picture of Analog/Digital Sound Encoding 65
   7.2 LP and Tape Technologies 68
   7.3 The Digital Approach and Sampling 70

8   Finite Fourier 75
   8.1 Finite Fourier Analysis 75
   8.2 Fast Fourier Transform (FFT) 78
      8.2.1 Fourier via Complex Numbers 78
      8.2.2 The FFT Algorithm 81
   8.3 Compression 82
   8.4 MP3, MP4, AIFF 84

9   Audio Effects 93
   9.1 Filters 93
   9.2 Equalizers 96
   9.3 Reverberation 98
   9.4 Time and Pitch Stretching 100

Part IV Musical Instrument Digital Interface (MIDI)

10 Western Notation and Performance 109
    10.1 Abstraction and Neumes 110
    10.2 Western Notation and Ambiguity 111

11 A Short History of MIDI 115

12 MIDI Networks 117
    12.1 Devices 117
    12.2 Ports and Connectors 118

13 Messages 121
    13.1 Anatomy 121
    13.2 Hierarchy 123

14 Standard MIDI Files 125
    14.1 Time 125
    14.2 Standard MIDI Files 125

Part V Software Environments

15 Denotators 131

16 Rubato 139
    16.1 Introduction 139
    16.2 Rubettes 139
    16.3 The Software Architecture 141

17 The BigBang Rubette 143

18 Max® 145
    18.1 Introduction 145
    18.2 Short History 146
    18.3 Max® Environments 147
    18.4 Some Technical Details 151
    18.5 Max®Artists 154

Part VI Global Music

19 Manifolds in Time and Space 159
    19.1 Time Hierarchies in Chopin's Op. 29 159
    19.2 Braxton's Cosmic Compositions 160

20 Music Transportation ... 163
- 20.1 Peer-to-Peer Networking ... 163
- 20.2 Downloads for Purchase ... 165
  - 20.2.1 A Simple Example of Encryption ... 167
  - 20.2.2 FairPlay: Fair or Unfair? ... 169
- 20.3 The Streaming Model ... 170
  - 20.3.1 Effects on Consumers and Industry ... 171

21 Cultural Music Translation ... 173
- 21.1 Mystery Child ... 173
- 21.2 Mazzola's and Armangil's Transcultural Morphing Software ... 174

22 New Means of Creation ... 177
- 22.1 The Synthesis Project on the Presto Software ... 177
- 22.2 Wolfram's Cellular Automata Music ... 178
- 22.3 Machover's Brain Opera ... 180
- 22.4 The VOCALOID™ Software ... 183
  - 22.4.1 Introducing VOCALOID™: History ... 183
  - 22.4.2 VOCALOID™ Technologies ... 183
- 22.5 The iPod and Tanaka's Malleable Mobile Music ... 185

References ... 187

Index ... 191

Part I

Introduction

1
General Introduction

Summary. This introduction gives a general orientation to the book's topic and philosophy.

$$- \Sigma -$$

Music technology deals with the knowledge and engines for musical creation and management. It includes mechanical and electromagnetic devices, such as musical instruments, electrical sound generators, and computer-oriented technology with its software innards. Despite the modern word "technology," music was from its very beginnings intimately related to the technological realms. Already the first instruments, such as the 50,000-year-old bone flute discovered in southern Germany, and the lyra of ancient Greek culture, are simple technological devices for the production of musical sounds. The Pythagorean school around 500 B.C. added a knowledge basis to its musical instruments, generating the far-reaching mathematical concept of a consonant interval in its metaphysical tetractys formula. This conceptual basis of music was also realized in other cultures, such as, prominently, bell tuning in China around 433 B.C.

In this sense, music has always been substantially intertwined with its technology. Music without technology would be a recipe without ever cooking, an abstract nonsense that does not have any substantial meaning. But we have to be aware that music deeply engages in a *balanced dialog* between knowledge and its technical realization. Musical creation vitally depends on lessons from acoustics and music theory about the construction of instrumental hardware. For example, the geometry of the piano keyboard is a direct consequence of pitch being the logarithm of frequency: octaves are all equidistant. Another example is the major scale's selection of pitches: the white keys. And vice versa: Music theory in the broadest sense of the word reflects the impact of instrumental constructions. For example, the penetration of sound spectra by electronic instruments has entailed the theory of spectral music with the development of theoretical concepts that include timbre and not only pitch and onset parameters [20].

Following this dialogical principle, after a short discussion of musical ontology in Part I, our presentation gets off the ground with the discussion of physical reality in Part II: acoustics. We introduce the most important descriptions of sound anatomy by Fourier's theory of partials, Chowning's frequency modulation, wavelets, and physical modeling. We then have a closer look at a selection of mechanical and electronic instruments. This part terminates with the presentation of the Euler space of pitch, together with its signification for tuning and contrapuntal interval categories.

Part III introduces the core technology of electromagnetic encoding of music, including a reasonably accessible introduction to the corresponding theory: finite Fourier analysis, the FFT (Fast Fourier Transform) algorithm, MP3 compression, filters, and time/pitch stretching. This is probably the most exciting part of the book as it is difficult to find easily accessible texts in the literature.

Part IV discusses the leading musical information exchange format: MIDI (Musical Instrument Digital Interface), the standard language of the music industry for communicating musical event commands among computers, synthesizers, and digital instruments. It is important to understand that MIDI, much like the traditional Western score notation, is a symbolic language that relates to acoustics only marginally.

Part V introduces two significant software environments of music technology: first, the universal concept architecture of denotators and forms (developed by Guerino Mazzola and his research groups since 1992) including its implementation in the software RUBATO® for analysis, composition, and performance, and second, the prominent Max® environment with its sophisticated arsenal of visual interface components for sound synthesis.

The last part VI is entitled "Global Music" and gives an overview of recent tendencies of musical globalization with transcultural composition software and Internet-based applications such as iTunes, Spotify, Todd Machover's Brain Opera, and other far-out innovations in music technology.

2
Ontology and Oniontology

Summary. This short chapter introduces the global architecture of ontology of music, which this book is going to use extensively.

$$-\Sigma-$$

This chapter is about ontology of music, including three dimensions: realities, semiotics, and communication. It also includes the extension of ontology to the fourth dimension of embodiment. We call this extension "oniontology" for reasons that will become evident soon.

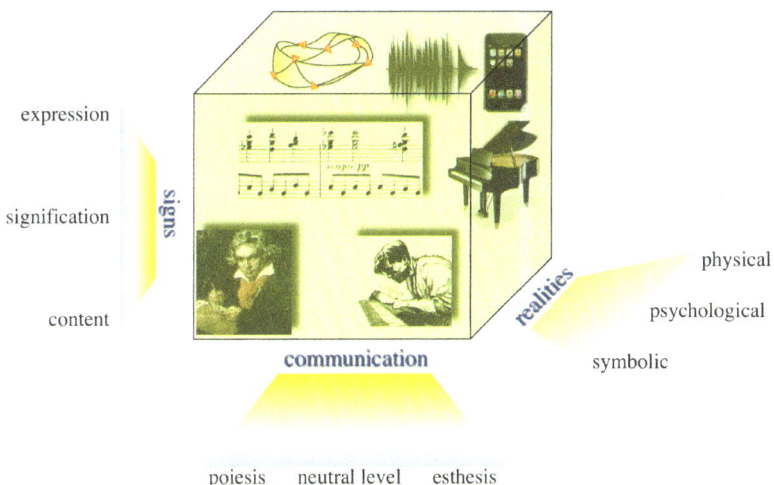

Fig. 2.1: The three-dimensional cube of musical ontology.

2.1 Ontology: Where, Why, and How

Ontology is the science of being. We are therefore discussing the ways of being that are shared by music. As shown in Figure 2.1, we view musical being as spanned by three 'dimensions', i.e., fundamental ways of being. The first one is the dimension of realities. Music has a threefold articulated reality: physics, psychology, and mentality. Mentality means that music has a symbolic reality, which it shares with mathematics. This answers the question of "where" music exists.

The second dimension, semiotics, specifies that musical being is also one of meaningful expression. Music is also an expressive entity. This answers the question of "why" music is so important: it creates meaningful expressions, the signs that point to contents.

The third dimension, communication, stresses the fact that music exists also as a shared being between a sender (usually the composer or musician), the message (typically the composition), and the receiver (the audience). Musical communication answers the question of "how" music exists.

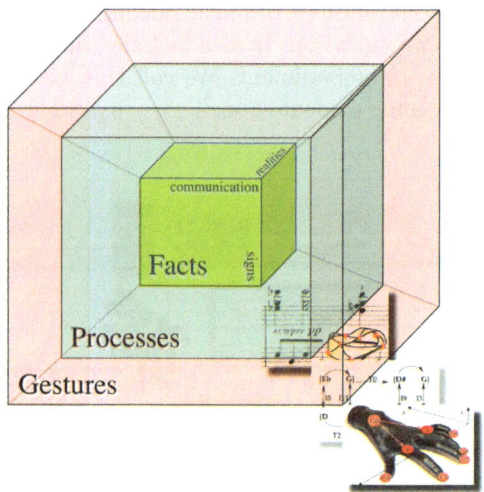

Fig. 2.2: The hypercube of musical oniontology.

2.2 Oniontology: Facts, Processes, and Gestures

Beyond the three dimensions of ontology, we have to be aware that music is not only a being that is built from facts and finished results. Music is strongly also processual, creative, and living in the very making of sounds. Musical performance is a typical essence of music that lives, especially in the realm of

2.2 Oniontology: Facts, Processes, and Gestures

improvisation, while being created. The fourth dimension, *embodiment*, deals with this aspect; it answers the question "how *to come into* being?" It is articulated in three values: facts, processes, and gestures. This fourth dimension of embodiment gives the cube of the three ontological dimensions a threefold aspect: ontology of facts, of processes, and of gestures. This four-dimensional display can be visualized as a threefold imbrication of the ontological cube, and this, as shown in Figure 2.2, turns out to be a threefold layering, similar to an onion. This is the reason why we coined this structure "oniontology"—it sounds funny, but it is an adequate terminology.

Part II

Acoustic Reality

3
Sound

Summary. In this chapter, we explain four essential aspects of sound. We begin by discussing the most basic aspect, the acoustic reality. This describes the production, propagation, and reception of sounds. The second aspect is sound anatomy, a mathematical model of an individual sound. The third aspect is the communicative dimension of sound as a message. The fourth aspect is the human anatomy of sound perception.

$$- \varSigma -$$

3.1 Acoustic Reality

Acoustics deals with the physics of sound. Sound is generated by a sound source, typically an instrument or human voice in musical contexts (see Figure 3.1). This sound source acts on the molecules of air and makes them produce a variable air pressure (around the normal pressure of 110.130 Pa, $Pa = Nm^{-2}$, at sea level) that propagates through space, is redirected by walls and objects, and then reaches the human ear, which is sensitive to such pressure variations. The auditory nerves in the cochlea in the inner ear then conduct the sound input to the auditory cortex and other sensitive brain centers, where different properties of the sound information are perceived and processed for the human cognition.

Let us first focus on the instrumental source structure providing the sound, i.e., on the complex trajectory of the sound from the source to the ear. The problem of controlling the trajectory is crucial. Historically, this relates to room acoustics: How can we shape a music hall in such a way that performed works can be heard in an optimal way? There are a number of technologies to split the trajectory into a multistep pathway. The sound can be piped into a microphone system and then distributed to a number of loudspeakers. This has been done not only for simple reasons of acoustical quality in music halls and dance clubs, but also to redefine the output in an artistic way, for example

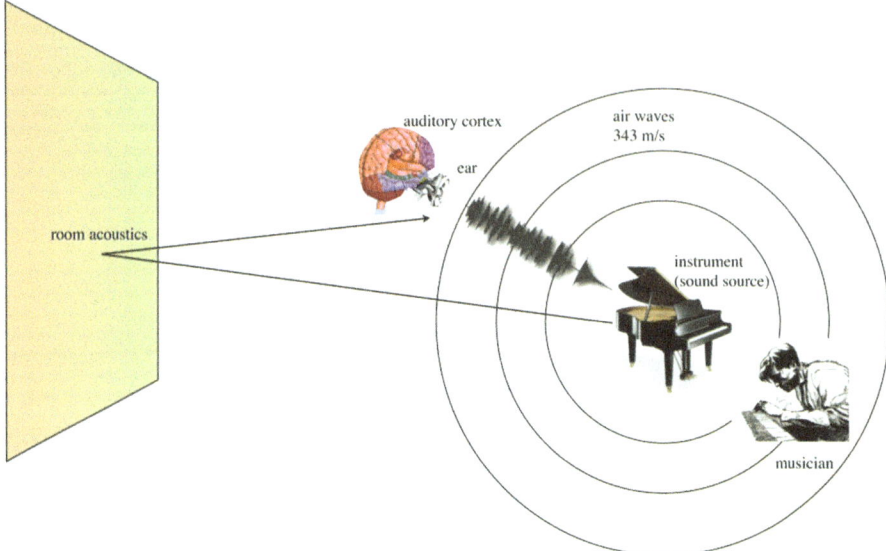

Fig. 3.1: Sound is generated by a sound source, typically an instrument or human voice in musical contexts. This sound source acts on the molecules of air and makes them produce a variable air pressure that propagates through space, is redirected by walls and objects, and then reaches the human ear, which is sensitive to such pressure variations. The auditory nerves in the cochlea in the inner ear then conduct the sound input to the auditory cortex and other sensitive brain centers, where different properties of the sound information are perceived and processed for the human cognition.

when a multiplicity of loudspeakers is part of the composition and may enable migration of sound among several loudspeakers.

Another interruption of the sound trajectory might be to save the sound on a tape or digital sound file. This relay station can then be used to reconstruct the sound and apply any kind of sound processing before forwarding the sound to the audience. Finally, following some ideas from the Venetian polychoral style of distributed choirs, one may also rethink the role of the music hall's space, extending use of the stage to the entire hall (including the audience space and even the music hall's bar) and setting up musicians to play anywhere.

3.2 Sound Anatomy

The structure of musical sound waves[1] is described in Figure 3.2. They consist of longitudinal waves, i.e., the pressure variation in the air moves in the

---

[1] Mathematically speaking, a wave is a function $w(x,t)$ of spatial position x and time t such that $w(x, t+P) = w(x - v \cdot P, t)$, i.e., the value at x after a period P of

direction of the sound propagation away from the sound source. If they carry the pitch, the variation of pressure is periodic, i.e., a snapshot of the pressure shows a regularly repeated pressure along the sound's expansion in space. The time period P of this regularity, as shown in Figure 3.2, defines the sound's frequency $f = 1/P$. For example, the chamber pitch a^1 has frequency 440 Hz (Hz (Hertz) is the frequency unit, i.e., periods per second). And going an octave higher means doubling the frequency; therefore, the octave a^2 above a^1 has frequency 880 Hz. Pitch is related to frequency f by the formula $Pitch(f) = \frac{1200}{log_{10}(2)} log_{10}(f) + C\ Hz$.

The wave's pressure amplitude A (relative to the normal pressure) is perceived as loudness. For example, the minimal loudness that can be perceived by the human ear at 1,000 Hz is $A_0 = 2.10^5\ Nm^{-2}$. Loudness for amplitude A is then defined by $l(A) = 20.\log_{10}(A/A_0)\ dB$, where the unit dB is *decibel*.

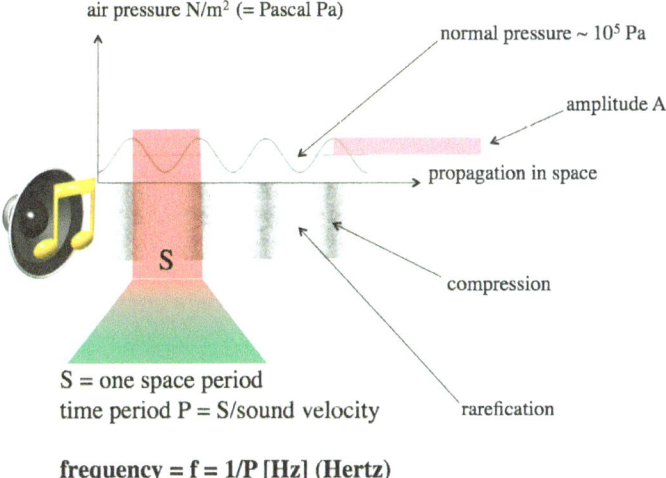

Fig. 3.2: Standard parameters of sound with frequency (pitch).

A sound is a wave, starting at an onset time, lasting a certain time (its duration), and during that time showing the amplitude and periodicity discussed above. Taking this kind of object as an element of music, however is a problem. To begin with, it is not clear why these attributes should hold during the entire sound. Why should amplitude be constant? A sound may well increase its loudness. And then, why is the pitch constant? What about glissandi? And finally, where do we find reference to the sound color, the instrumental characteristic?

Onset, duration, loudness, pitch, instrument name—all of this is described in classical score notation. But only when comparing these forms to the real

---

time is the same as the value at the position traversed by the wave at the earlier time t.

sound events does one become aware that one can create richer musical compositions. We shall come back to such options in the discussion of sound synthesis methods, such as Fourier, FM, wavelets, and physical modeling.

3.3 The Communicative Dimension of Sound

Summary. Apart from the acoustic elements that comprise the nature of the interaction between sound and the human ear, there are dimensions in the nature of sound that profile its ontology. These include: *Reality, Communication,* and *Semiosis*. In this section we are going to discuss the communicative dimension of sound.

$$- \Sigma -$$

3.3.1 Poiesis, Neutral Level, Esthesis

Communication is the second dimension that defines the topography of sound. The semiologist Jean Molino divides the communication process into three realms: *creator, work,* and *listener.* The creator or broadcaster produces the sound, which enters the physical reality and reaches the receiver, who is the listener. Communication describes the transition procedures between the three different realms.

In an effort to study and analyse the aforementioned realms, three corresponding concepts were developed by Molino: *poiesis, neutral level,* and *esthesis,* see Figures 3.3, 3.4, and 3.5. *Poiesis* comes from the Greek word $\pi o \iota \epsilon \iota \omega$, which corresponds to the processing of elements that the creator encompasses in the creation of sound. *Poiesis* includes every exhibition of creative activity that involves either composition or improvisation. Furthermore, *poiesis* is strictly focused on the making of sound, not on the intended object itself.

The outcome of the creative process is the *work*, whicht corresponds to the concept of the *neutral level*. The *work* subsumes the entirety of stimuli in the creator's process. Therefore, we are dealing with the concept of a discourse that is "independent of the selection of the tools and is strictly oriented towards the given work" [37, Section 2.2.2]. Moreover, it opposes any type of external valuation of the work, whether historical or social.

The *listener* is the receiver and interpreter of the *work*. The listener and the creator have different points of view regarding the *work*; the creator produces the *work* while the listener perceives an already existing work. The discourse on the experience of the listener is called Esthesis, originating from the Greek word $\alpha \iota \sigma \theta \eta \sigma \iota \varsigma$, which means perception. The listener receives and valuates the work from his/her own individual point of view. This act of singular valuation of the piece carries the same energy and significance as does the creator's labour. The distinction between the poietic and the esthetic realms is clear. Rarely do we witness the faithful transfer of instances from the poietic to the esthetic level. A good example is free jazz: The performers compose on the spot and interact with each other, an intact compositional outcome. Nevertheless, it is common that concepts from the poietic realm are falsely used at the esthesic level, confusing the distinction between the productive and the perceptive processes, obscuring the creator's and the listener's different positions in the timeline of the work's realization. Poietic and esthesic levels meet only when the compositional process is approached in retrograde; in *retrograde poiesis* the creator is the first observer of the work and, therefore, retrograde poiesis is incorporated in esthesis.

Fig. 3.3: Poiesis   Fig. 3.4: Neutral level   Fig. 3.5: Esthesis

3.4 Hearing with Ear and Brain

Summary. As we have learned, sound is created by differences in air pressure, by waves traveling through air. However, our perception of sound is not directly related to differences in pressure. Our brain cannot receive this information directly. Instead, the human body uses a series of mechanisms to codify this information, provide the code to neuronal cells, and send this neural code to the brain.

$$-\Sigma-$$

Sound is created by differences in pressure, by waves traveling through air. However, our brain cannot receive this information directly. Instead, the human body uses a series of mechanisms to codify this information, provide the code to neuronal cells and send this neural code to the brain.

Your sense of hearing is the result of this translation and transmission of acoustical information. How does the body do this? The ear has three main parts: the outer ear, the middle ear, and the inner ear.

The outer ear receives incoming sound and vibrates based on the differences in pressure present in the auditory canal. The eardrum separates the outer ear from the middle ear. The vibration of the eardrum is passed to three bones in the middle ear, referred to collectively as the *ossicles*. The ossicles are the smallest bones in the human body. Due to their size, they are highly sensitive to vibration. The inner ear, called the *cochlea*, is filled with fluid. In order for the ossicles to create a pressure difference that propagates in the new fluid medium, they concentrate the force of vibrations at the *oval window*, a bridge between the middle ear and the cochlea (see Figure 3.6) directly connected to the stapes. In addition to the oval window, the cochlea has a *round window*, which dampens the vibrations that have traveled through the cochlea. Without the *round window*, the residual vibration would cause us to hear everything twice, like an echo. This would inundate the brain with an excessive amount of auditory information.

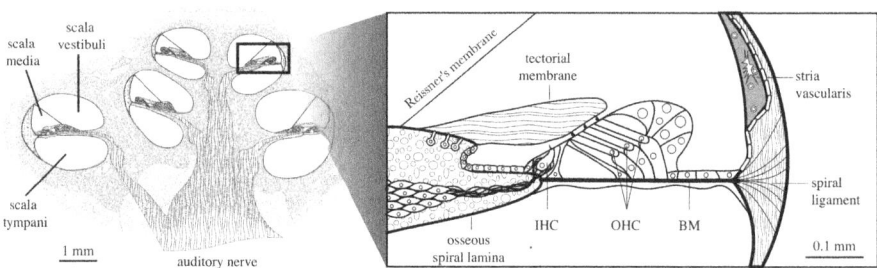

Fig. 3.6: Section of Cochlea. IHC: Inner Hair Cells, OHC: Outer Hair Cells, BM: Basilar Membrane. The *scala vestibuli* receives the vibrations of the oval window.

The cochlea is shaped like a hollow spiral. Inside this 'snail's shell' is the *basilar membrane*, which contains 20,000 neural hair cells responsible for coding vibrations in the fluid into electrical signals.

Now that sound has been translated from pressure differences in the air into electrical impulses in the cochlea's hair cells, what happens next? In short, the body's equivalent of electrical wiring takes on the responsibility of getting that information to the brain, where it can be processed. This network is known as the , and it is based on electrical and chemical communication between neurons.

Once this information is received by the neurons, it is sent to the auditory cortex in the temporal lobe of the brain. The auditory cortex translates this information into what we perceive as pitch. After this point, it is unclear exactly how the brain is able to translate pitch, rhythm, articulation, and other aspects of music into a profound cognitive and emotional experience. However, it is clear that listening to music affects many parts of the brain, influencing parts of the brain responsible for pleasure (i.e., the nucleus accumbens) and emotion (e.g., the amygdala and hippocampus).

The amazing neurological influence of music is not restricted to music listening, but extends to music playing and learning. In fact, learning an instrument produces significant changes in the brain that can change one's behavior and ability. For example, Oechslin et al. [48] found that individuals who participated in music lessons for a number of years were found to have a greater hippocampal volume (meaning they experience music emotion at a higher level) than those who did not take lessons. Oechslin and his colleagues also found that individuals with higher hippocampal volume tested higher on creative aptitude tests. The researchers attributed increased creative aptitude to the increased hippocampal volume.

Music is not only a fascinating topic of neuroscientific study, but (due to its ability to activate a wide variety of brain areas) it also plays an important role in teasing apart different parts of neuroscience and making new discoveries. Music is not only a sensationally rich stimulus. It has an astonishing ability to access and trigger our thoughts and emotions. Given its omnipresence in the brain, surely we can say that music is an intrinsic component of the human experience.

4
Standard Sound Synthesis

Summary. In this chapter we immerse ourselves in the core elements that constitute a sound; we discuss waves, their specifics, and the frequency content of a sound. Moreover, we elaborate on the sinusoidal functions and the processes that characterize them and have revolutionized modern technology, culminating in Fourier's Theorem.

$$- \Sigma -$$

4.1 Fourier Theory

4.1.1 Fourier's Theorem

Using very simple devices that act like 'atomic instruments', it is possible to produce a complex sound that not only shares the usual musical sound parameters, such as pitch, onset, duration, and loudness, but also instrumental timbre or sound color.

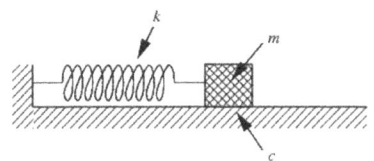

Fig. 4.1: Fourier's atomic instrument.

Among these attempts, Joseph Fourier's theorem, discovered in research into heat conductance around 1800, provides a first model. The 'atomic instruments' he uses are vibrating masses m attached to a fixed point at distance c by a spring with spring force constant k, and moving along a fixed line, see Figure 4.1. Newton's differential equation (Newton's Second Law) for the force

4 Standard Sound Synthesis

acting upon m is $m\frac{d^2c}{dt^2} = -kc$, i.e., mass (m) times acceleration ($\frac{d^2c}{dt^2}$) is equal to the force of the spring. The solution of this vibrating system is a sinusoidal function $c(t) = (\frac{k}{m})sin(t)$.

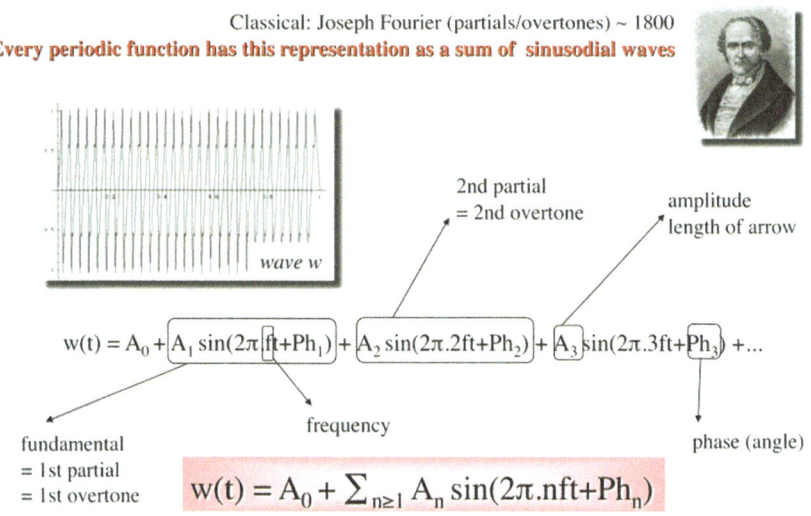

Fig. 4.2: Joseph Fourier's theorem, discovered in research about heat conductance around 1800, provides a first model of sound as composed of 'atomic' sound shapes: sinusoidal functions.

Recall that the pitch of a sounding periodic air pressure vibration $w(t)$ is proportional to the logarithm of its frequency $f = \frac{1}{Period}$. Fourier's theorem then states that $w(t)$ can be expressed in a unique way as a sum of sinusoidal functions, or, so to speak, as a sum of those atomic instruments given by masses and springs (see Figure 4.2). Uniqueness means that for the given frequency f of $w(t)$, the *amplitudes* $A_0, A_1, A_2, \ldots$ and the *phases* $Ph_1, Ph_2, \ldots$ are uniquely determined (and called the amplitude and phase spectrum, respectively). The nth component function $A_n \sin(2\pi n f t + Ph_n)$ is called the nth *partial* or *overtone* of $w(t)$.

Let us describe how the Fourier theorem is related to realistic sounds. We can see that the *wave* function $w(t)$ (at a fixed spatial position) is anything but natural. In reality no such infinitely lasting regular air vibration can occur. The relationship to realistic sounds can be seen in Figure 4.3. If a singer sings "laaaa" at a determined pitch, the pressure variation around the mean pressure looks like a bundle, as shown at the top of Figure 4.3. The bundle can be described by its envelope, i.e., the locally maximal pressure variations (shown at the left bottom), and by a periodic excitation of pressure, limited by the envelope, and shown at the right bottom.

4.1 Fourier Theory   21

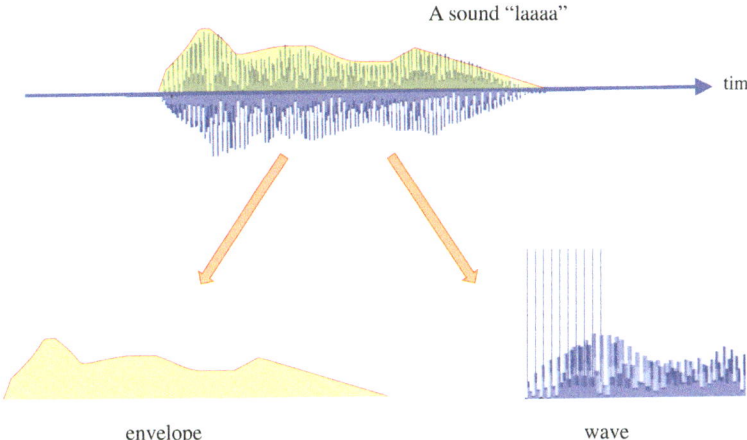

Fig. 4.3: When a singer sings the syllable "laaaa" at a determined pitch, the pressure variation around the mean pressure looks like a bundle as shown on top. The bundle can be described by its envelope, i.e., the locally maximal pressure variations (shown at the left bottom), and by a periodic excitation of pressure, limited by the envelope, and shown to the right bottom.

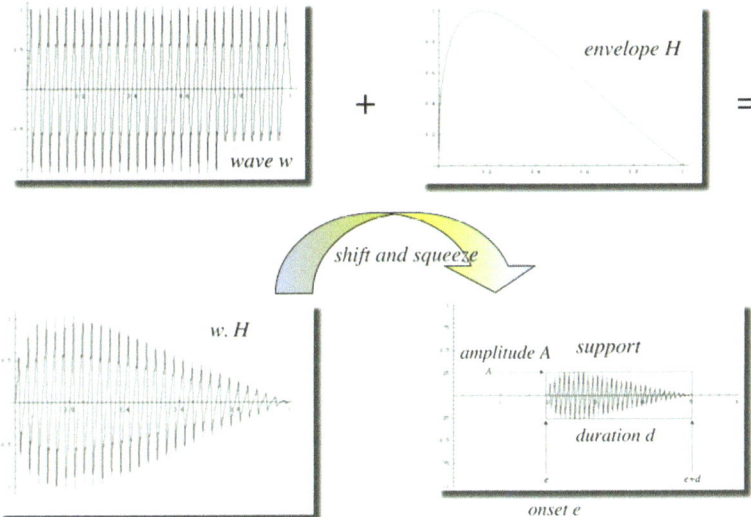

Fig. 4.4: The combination of envelope and periodic function yields the realistic sound.

This combination of envelope and periodic function yields the realistic sound. In a systematic building process, as shown in Figure 4.4, this means that the periodic wave $w(t)$ is superimposed with the envelope H (normed to duration and maximal excitation both = 1), and then shifted and squeezed by a *support* to yield theonset, duration, and amplitude of the real sound.

In technological modeling of this building process, the envelope is often represented as a very simple shape, following the ADSR (Attack Decay Sustain Release) model, as shown in Figure 4.5. Attack is characterized by the time that passes while the spectrum of the sound is formed. The attack of a sound is a major indicator of the instrument used. For example, a guitar might sound like a bell without the guitar's distinctive attack. Sustain refers to the steady state of sound when it reaches its highest energy. Finally, decay is defined by the rate at which the intensity of the sound disappears to silence. Every sound has a particular pattern of attack, sustain, and decay; therefore the envelope of a sound is characteristic of the sound.

Although this seems to describe realistic sounds quite faithfully, it turns out that instrumental sounds are more complex in that the partials are given independent amplitude envelopes that enable the sound to have variable overtones as it evolves in time. Figure 4.6 shows such a situation for a trumpet sound. The graphic displays envelopes for each partial, and we see that higher partials have lower and shorter envelopes, meaning that the amplitudes of these partials contribute for only a short time and to a lower degree to the overall sound. The display of these temporally variable overtone envelopes is called a *chronospectrum*.

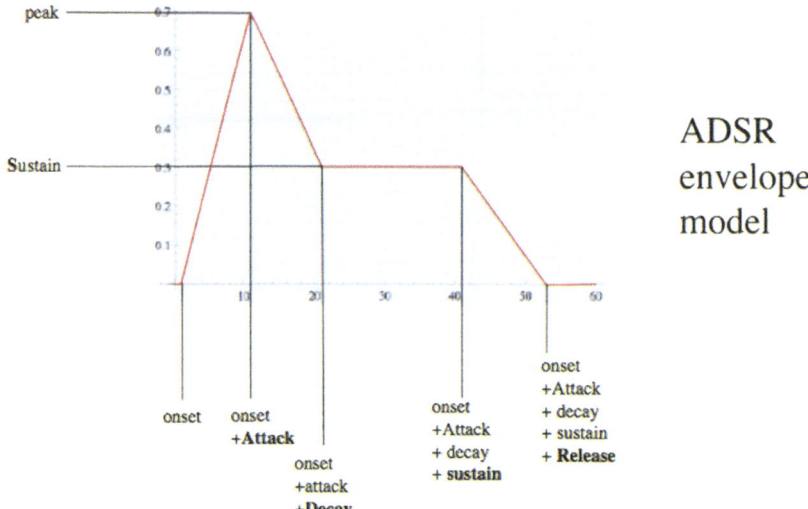

Fig. 4.5: The envelope is often represented as a very simple shape, following the so-called ADSR (Attack Decay Sustain Release) model.

Fourier's theorem and the description of sounds by overtones are firm scientific facts, as a mathematical theorem cannot be deemed false once its proof is correct. Why should such a fact be a problem for musical creativity? To begin with, its facticity seems to condemn everybody to its unrestricted acceptance. It is an eternal truth. And can't an overtone-related analysis—although not precisely true [37, Appendix B.1]—in the inner ear cochlea's Corti organ be taken as a strong argument for the universal presence of the Fourier decomposition of a sound?

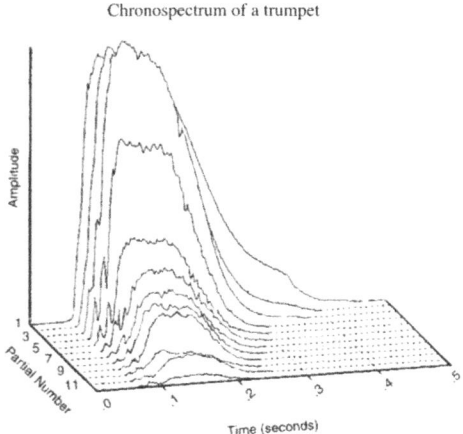

Fig. 4.6: A trumpet sound. The graphic displays envelopes for each partial, and we see that higher partials have lower and shorter envelopes, meaning that the amplitudes of these partials contribute only a short time and to a lower degree to the overall sound. The display of these temporally variable overtone envelopes is called *chronospectrum*.

Let us think about the justification of this theorem: We are analyzing a sound, and Fourier tells us how to do so. But analysis is not synthesis, and after all, composers are interested in synthesis rather than analysis. The point here is whether this formula is the end of an insight or the beginning thereof. If it is just about understanding a periodic function, it is the end,[1] but if it is about putting together atomic instruments (the sinusoidal functions) to form a compound instrument, it is a new thing. We can construct instruments by adding sinusoidal functions that are provided with their individual amplitude envelopes. And we do not have to add sinusoidal functions with frequencies that are multiples of a fundamental frequency. This latter generalization is called *inharmonic partials*. Summarizing, there is no need to follow the original formula, but we can generalize it and generate compound sounds with the

---

[1] Well, not really, since there are infinitely many different systems of non-sinusoidal 'atomic sounds' that may also model complex sounds.

original building rule, but in a more flexible approach. This is precisely what spectral music composers discovered [20]. They were interested in composing sounds, not abstract symbols. And the synthetic interpretation of Fourier's formula was the technical tool they used to shape the sound's anatomy as a living body.

4.2 Simple Waves, Spectra, Noise, and Envelopes

Sound is the vibration of molecules of a certain transmission medium—such as air—transmitted in the form of a wave. The intrinsic characteristics of the motion and nature of a wave are responsible for the different attributes of sound. For example, Figure 4.7 shows the snapshot of a *basic sinusoidal wave*, sound example: sine.mp3.

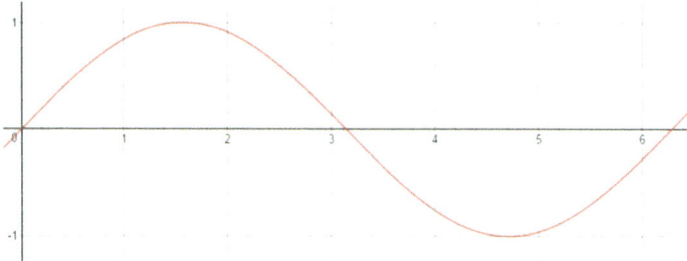

Fig. 4.7: A simple wave.

According to Fourier theory, a sound can be represented by its *spectrum*. A sound spectrum is graphically represented by the frequency content of a sound against amplitude of energy (square of amplitude) of each partial, see Figure 4.8.

The spectrum of a sound offers significant insight into its characteristics. It is displayed in a graphic representation of amplitude (vertical axis, measured in decibels (dB)) or energy (square of amplitude), and frequency (horizontal axis, measured in vibrations per second or Hertz (Hz)). The harmonics of a sound appear as vertical lines in the sound spectrum, see Figure 4.9.

The spectrum of a sound can be modified using *filters* (see Section 9.1). These effects have an impact on the original interrelation between amplitude and frequency, therefore producing a different signal.

Similarly, a *spectrogram* is a three-dimensional representation involving amplitude, frequency, and their variation in time. Time is displayed on the horizontal axis, frequency on the vertical axis, and amplitude is manifested through the intensity of colors, see Figure 4.10. Since spectrograms show the different frequencies that a sound is made of, they are an invaluable tool in the study of timbre.

4.2 Simple Waves, Spectra, Noise, and Envelopes   25

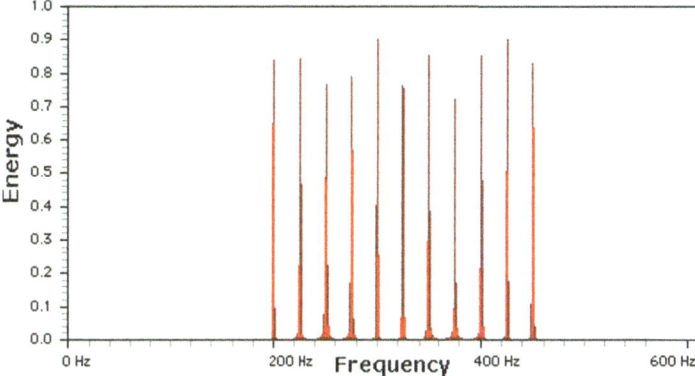

Fig. 4.8: Sound Spectrum: the amplitude is measured in decibels (dB) and the frequency is measured in vibrations per second or Hertz (Hz). Here the energy, square of amplitude, is shown.

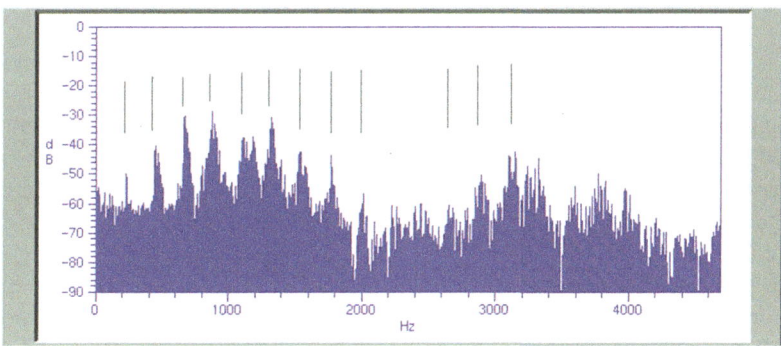

Fig. 4.9: The spectrum displays the amplitudes of the sound's harmonics as peaks.

Noise is considered to be any annoying or unwanted sound. The difference to wanted sounds is based upon the subjective perception of each person; an agreeable sound for some people can be perceived as noise by others. But in acoustics, sound and noise are distinguishable: Sound has a discrete set of frequency components, its fundamental and overtones. Noise, on the other hand, shows a continuous spectrum (see Section 9.1). There are many types of noise, and all have been named after colors; the color of a noise describes the distribution of frequency intensities in a wide range of the audible spectrum.

For example, *white noise* (Figure 4.11, sound example: `wnoise.mp3`) contains equal energy in all frequencies in analogy to white light, which contains almost equal energy in all frequencies of the visible spectrum. *Pink noise* (Figure 4.12, sound example: `pnoise.mp3`) is characterized by a 3 decibels per octave decrease in energy for increasing frequency. This decrease rate corresponds to the decrease rate of non-electronic instruments. For this reason pink

26   4 Standard Sound Synthesis

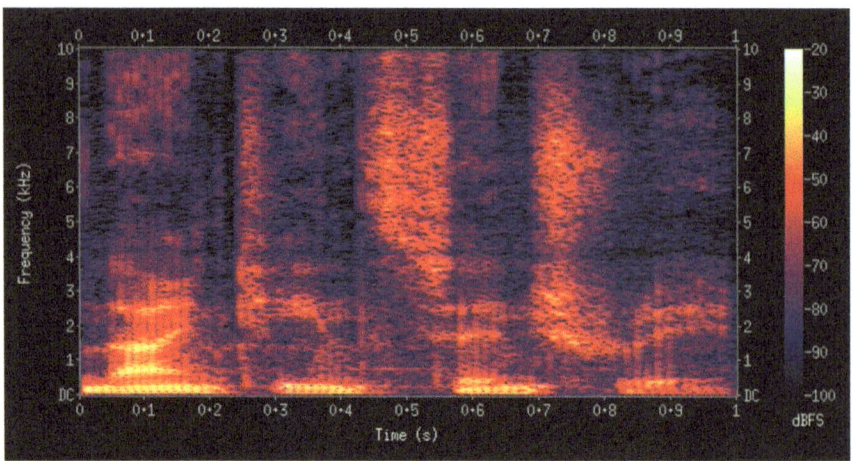

Fig. 4.10: A spectrogram.

noise is used for checking the acoustic features of auditoriums and other sonic spaces. *Grey noise* (Figure 4.13, sound example: gnoise.mp3) is characterized by equal loudness for all frequencies. Since the human ear is more sensitive to a certain frequency range, we tend to perceive certain frequencies more intensely than others. What grey noise does is to balance this difference by adding energy to the high and low range of frequencies and subtracting energy from the middle range, where hearing is easier. In contrast, *blue noise* (Figure 4.14, sound example: bnoise.mp3) involves a 3 decibels per octave increase in energy for increasing frequency, while *violet noise* increases 6 decibels per octave for frequency increase. Last but not least, in *violet noise, brown or red noise* (Figure 4.15, sound example: rnoise.mp3), the lower the frequency, the higher the intensity.

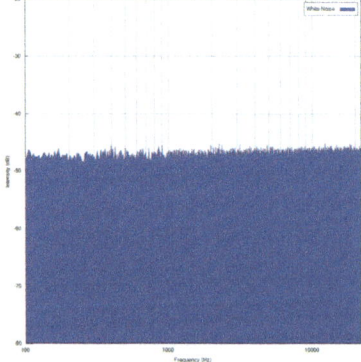

Fig. 4.11: The white noise spectrum.

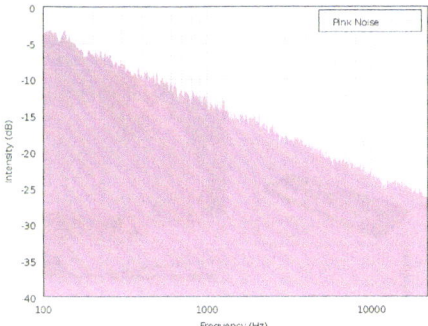

Fig. 4.12: The pink noise spectrum.   Fig. 4.13: The grey noise spectrum.

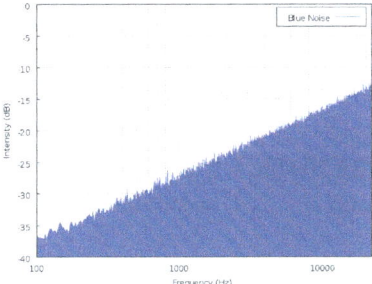 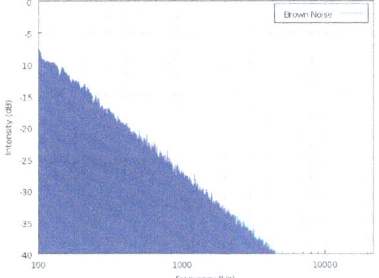

Fig. 4.14: The blue noise spectrum.   Fig. 4.15: The brown noise spectrum.

4.3 Frequency Modulation

Fourier's formula has not only provoked more general compositional approaches, such as spectral music, but also a more realistic sound representation for the technology of sound synthesis.

The first innovation for this purpose was introduced by John Chowning in 1967 at Stanford University and was called FM for "Frequency Modulation." His invention was patented and had a remarkable commercial success with Yamaha's DX7 synthesizer from 1983 (see Figure 4.16).

The creative idea was again to review the critical concept in Fourier's theory. Fourier's method of sound synthesis requires a huge number of overtones and of corresponding envelopes. This was critical in the context of sound synthesis technology. However, Chowning questioned its arithmetical construction. This was a delicate approach since addition of sinusoidal functions for the

4 Standard Sound Synthesis

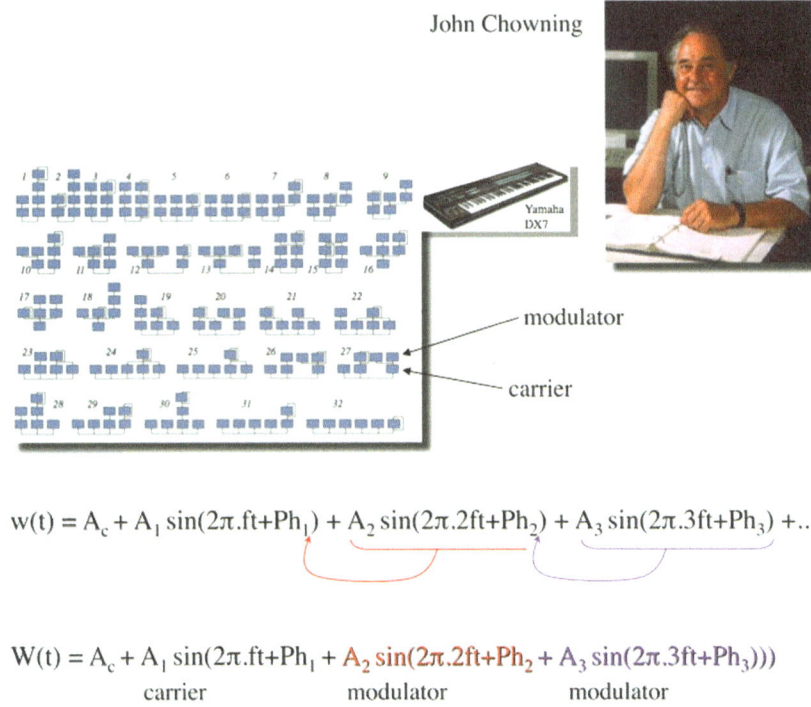

$$w(t) = A_c + A_1 \sin(2\pi.ft+Ph_1) + A_2 \sin(2\pi.2ft+Ph_2) + A_3 \sin(2\pi.3ft+Ph_3) + \ldots$$

$$W(t) = A_c + A_1 \sin(2\pi.ft+Ph_1 + A_2 \sin(2\pi.2ft+Ph_2 + A_3 \sin(2\pi.3ft+Ph_3)))$$
$$\quad\quad\quad\text{carrier} \quad\quad\quad\quad\quad \text{modulator} \quad\quad\quad\quad \text{modulator}$$

Fig. 4.16: John Chowning's invention of "Frequency Modulation" (FM) in 1967 at Stanford University (Credit: Photo of John Chowning by Fabian Bachrach for the Computer History Museum).

partials is difficult to question. Should one just negate addition, and then how should it be replaced or embedded by another operation? Should one multiply sinusoidal waves? Or apply any other operation? There are infinitely many such candidates, mostly unattractive because of their mathematical properties. And the problem was not only addition as an operation but also the number of summands necessary to produce interesting sounds.

Chowning's idea was highly creative in that he succeeded in replacing addition with another operation, which also allowed him to massively reduce the number of sinusoidal components. As seen from the list of components used in the Yamaha DX7, only six components are used in this technology. See [16] for sound examples. How can we build complex sounds with only six overtones?

The original formula for a wave $w(t)$, as shown in Figure 4.16, takes a fundamental wave and changes its values by the values of a second overtone, and then this sum has values that are in turn changed by the values of a third overtone, etc. Chowning's idea was to have the higher overtones' influence not the values of the preceding overtones, but on their time arguments. Instead of having time in a linear expression $2\pi n f t + Ph_n$, Chowning allowed a nonlinear

4.3 Frequency Modulation

correction of the time argument, the nonlinearity being defined by the overtone function. This means that instead of adding to values, his new synthesis formula adds to arguments. This was a revolutionary idea. The result is shown in the formula $W(t)$ in Figure 4.16, where the second overtone value is added to the argument of the first overtone, and where the third overtone value is added to the argument of the second overtone. The overtone distortion of the time argument is called a modulator, and it in facts modulates the frequency, in the sense of a time function distortion of the frequency-related linear argument, see also Figure 4.17.

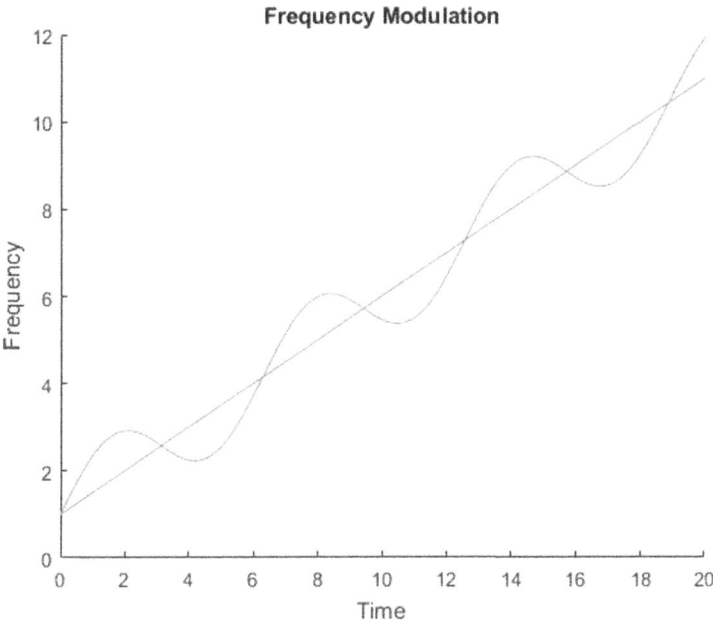

Fig. 4.17: Frequency modulation changes the frequencies of partials by adding "modulating" sinusoidal deviations to the given frequency argument.

In Chowning's approach, it is also permitted to add several overtone distortions to the linear time argument. The resulting architecture shows a number of sinusoidal functions (called *carriers*), which are altered by other sinusoidal functions (called *modulators*). Yamaha calls such an architecture an "algorithm." Yamaha's DX7 displays 32 such algorithms. Some of them are even their own modulators. It is possible to realize this with the technology since output and input values of these sinusoidal 'boxes' are shifted by elementary time units.

4.4 Wavelets

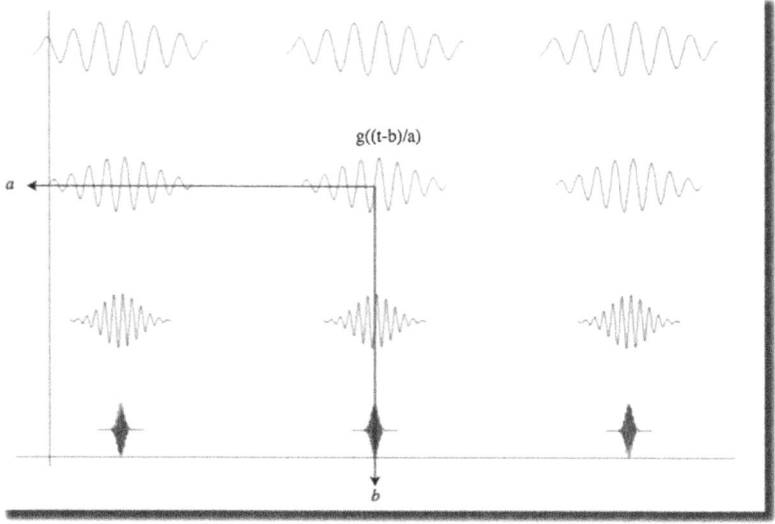

$$g(t) = C.\exp(-t^2/2).\cos(wt)$$
$$g_{a,b}(t) = g((t-b)/a)$$

Fig. 4.18: A wavelet function $g(t)$ of time. It is a sinusoidal function $\cos(wt)$ that has been modified by the envelope function $\exp(-t^2/2)$.

Although FM synthesis was a huge step in the reduction of sinusoidal components and the richness of sound colors realized in consumer technology, the FM formula is still far from realistic. Because it is infinite like Fourier's formula, the sinusoidal components are not limited-time functions. There is a canonical way to correct this, namely by envelopes. And this is exactly what was done in the approach of wavelet functions. Figure 4.18 shows such a wavelet function $g(t)$ of time. It is a sinusoidal function $\cos(wt)$ that has been modified by the envelope function $\exp(-t^2/2)$. Although this is still an infinitely extended function, its values converge quickly to zero by the exponential envelope. In wavelet theory, there are many such wavelet functions, but all have the property of showing the shape of a short wave 'package,' and this is why such functions are called *wavelets*. They play the role of sound atoms, much like sinusoidal waves in Fourier theory. Whereas sinusoidal waves are modulated by amplitude, frequency, and phase, wavelets are modulated by a time delay argument b and an expansion argument a to yield the deformed wavelets

$g_{a,b}(t) = g((t-b)/a)$. Figure 4.18 shows a number of such deformations that are placed in the plane of a and b.

The creative component of this theory is to take one single sinusoidal wave 'atom' $g(t)$, limit it by a strong envelope, and vary its time position and duration according to a whole two-dimensional parameter set (a, b). The question then is whether this two-dimensional variety of deformed and shifted wavelet atoms is capable of representing any (!) given sound function $s(t)$. The answer is positive, and it can be stated as follows: Define a coefficient function

$$S(b, a) = (\frac{1}{\sqrt{a}}) \int \bar{g}_{a,b}(t) \cdot s(t) dt,$$

where $\bar{g}_{a,b}(t)$ is the complex conjugate of $g_{a,b}(t)$. This function replaces the Fourier coefficient calculation. So we have an entire two-dimensional function, not only the discrete set of amplitudes and phases of the Fourier formula. But as with Fourier, this new function enables the reconstruction of the original sound wave by the following formula:

$$s(t) = Q_g \int g_{a,b}(t) \cdot S(b, a) da db,$$

where Q_g is a constant depending on the chosen wavelet type g. An example of the coefficient function is given in Figure 4.19, see also [25]. Here the grey value represents the numerical value of the coefficient function. The octaves in this musical example are visible as dark coefficient values.

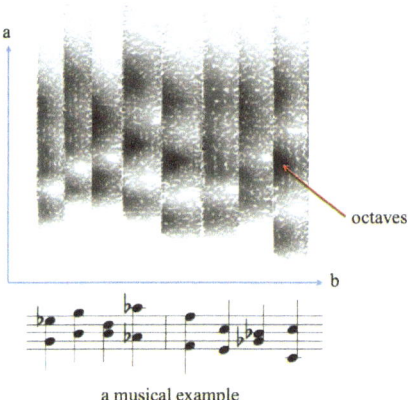

Fig. 4.19: An example of the coefficient function. The gray value represents the numerical value of the coefficient function. The octaves in this musical example are visible as dark coefficient values.

4.5 Physical Modeling

The last example of refined sound modeling is a radical change of paradigm. In Fourier's theorem, Chowning's FM, and the wavelet methods, we always dealt with sound representation. We have to recall however that instrumental technology has always been a radical answer to the question of sound synthesis. It is an answer of sound production, as opposed to the more or less abstract representation by mathematical formulas.

With the development of powerful personal computers, the instrumental approach has been recreated on the level of software simulation of entire instruments. It has become realistic to not only create an 'instrument' built from a series of sinusoidal waves a la Fourier, but also to simulate complex physical configurations on fast computers. This approach has three styles, namely:

- *The Mass-Spring Model*: The software simulates classical systems of point masses that are connected by springs and then outputs the sound wave resulting from the movement of such mechanical systems. A famous example is *Chordis Anima* developed by the Institute ACROE Grenoble, which was presented at the ICMC (International Computer Music Conference) in 1994.
- *Modal Synthesis*: Modeling of different sinusoidal components that are experimentally determined, e.g. *Modalys-Mosaic* at IRCAM (Institut de Recherche et Coordination Acoustique/Musique) [46].
- *Waveguide*: It models propagation of waves in a medium, such as air or string. This is implemented, for example, at Stanford's CCRMA (Center for Computer Research in Music and Acoustics), and a commercial version is realized on YAMAHA's VL-1.

Physical modeling is not just a simulation of physical instruments in software environments. By the generic character of such software, the parameters of the physical instruments, such as string length, diameter, material elasticity, material density, and so forth, can be chosen without any reference to realistic numbers. For example, physical modeling can simulate a violin with strings of glass and extending to several miles in length. So the software simulation in physical modeling now enables opening all the creative limitations of realistic physical parameters. It is like a reconstruction of physical devices with fictituous parameters.

A fascinating example of such a simulation has been implemented by Perry Cook in his singer synthesis software [11]. The software's functional units, simulating a human singer's physiology with throat, nose, and lips, is shown in Figure 4.20.

This seems to be a huge extension of instrumental technology, but one should observe that physical modeling deals only with the instrument as such, not with the instrument as an interface between human gestures and sounding output. In other words, the entire virtuosity of musicians on their instruments must be reconsidered when building interfaces capable of letting humans play

4.5 Physical Modeling   33

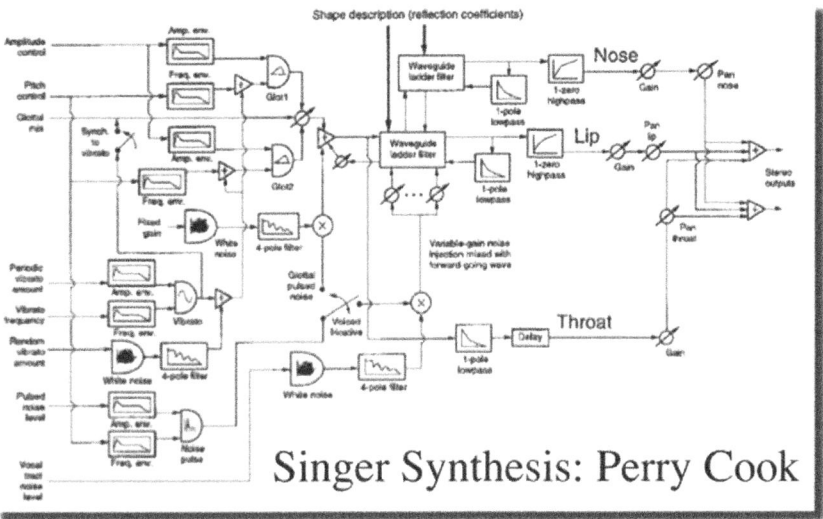

Fig. 4.20: Perry Cook's singer synthesis software.

such physically modeled sound generators. This means that creativity can, on the one hand, expand one concept—in this case, that of a musical instrument as a sound generator—but on the other hand also create new problems for the management of such expanded concepts, here the musician's interaction with these new instrumental devices.

To put it simply: It is not sufficient to create a new fancy instrument by physical modeling, but you also have to consider how to play the instrument. Moreover, to preserve the information about your creation, you would have to develop a new appropriate notation (analogous to the traditional Western notation, for example) and then also an educational system that teaches musicians how to play such notated information.

5
Musical Instruments

Summary. This chapter reviews a number of important mechanical and electronic instruments. See [19] for a detailed discussion of the physics of musical instruments.

$$- \Sigma -$$

5.1 Classification of Instruments

There have been many attempts to classify instruments throughout the centuries. Many of these classification systems rely on the pitch range of instruments, their historical relations, or common usage. The Hornbostel-Sachs system is favored in this book because of its choice to classify all instruments by the means of sound production, regardless of origin. This system is ideal for our discussion of their differing acoustic properties because it can adapt to almost any instrument, and closely related instruments will have similar acoustic properties. The Hornbostel-Sachs system divides instruments into five main categories:

- **Idiophones**—Instruments which produce sound when the entire body of the instrument vibrates. These include mallet percussion instruments, claves, cymbals, and triangles.
- **Membranophones**—Instruments with a stretched membrane that is responsible for producing the sound. This includes any drum.
- **Chordophones**—Instruments where sound is produced by the vibration of any number of strings stretched between two points. This includes pianos, members of the violin family, and guitars.
- **Aerophones**—Instruments where sound is produced by vibrating air within the instrument. These include woodwind and brass instruments.
- **Electrophones**—Instruments which have electrical action, electrical amplification, or where sound is produced by electrical means. This includes

5.2 Flutes

In aerophones (wind instruments) sound is produced by creating standing waves within a tube. High-pressure air travels through the instrument, exits through an opening (bell/tone hole/leak) and creates an inverse pressure wave, which travels back to the mouthpiece. The production of the wave varies with the type of instrument. Flutes, whether they are transverse or end-blown, create pressure waves by dividing an air stream between the inside and outside of the instrument. When the air restricted to the body of the flute has a higher pressure than the secondary stream, the pressure wave forces more air to the outer path. This distribution is quickly changed when the pressure imbalance reverses, and a regular standing wave is formed (see Figure 5.1).

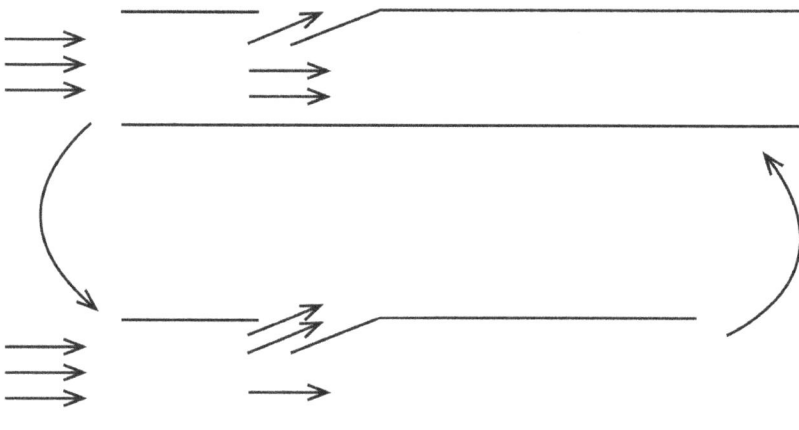

Fig. 5.1: The flow of air through a recorder mouthpiece. The division of the air stream rapidly fluctuates between the chamber of the instrument and the exterior.

The pitch produced depends on the length of the instrument. The flute is acoustically represented as a cylindrical pipe with two open ends; one at the end of the instrument and the other at the opening where the air enters the pipe. Pressure waves resonate in the chamber and escape at both ends of the pipe. Sound escapes the chamber at twice the rate it would for a pipe with only one opening. It's important to note that because there's a second opening at the head of the instrument, the standing waves of flutes can be extreme or neutral pressure at *both* ends simultaneously. Thus, fundamental standing waves in a flute are best represented by half of a simple sine curve with a periodicity that corresponds to the length of the pipe.(see Figure 5.2) Harmonic frequencies of the pipe exist at integer multiples of the fundamental standing wave's frequency, see Figure 5.3.

Fig. 5.2: A simple representation of the standing wave of an open tube.

Flute players frequently produce pitches on their instrument by "overblowing." By forcing more air through the instrument, a player can produce pitches in the harmonic series above the fundamental note. The greater the force of the breath, the higher the partial produced.

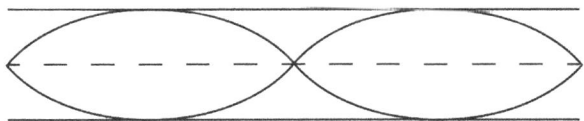

Fig. 5.3: A natural harmonic of an open tube. Note that the frequency is exactly twice the fundamental.

5.3 Reed Instruments

Reed instruments create pressure waves through the use of the reed as a gate. As a player blows into an instrument, the pressure inside the mouth pushes the reed towards the mouthpiece. As the reed bends, the force keeping it rigid increases. Eventually, the rigidity will overpower the player's air pressure, and the reed will snap back into place and the pressure is released into the body of the instrument. The process repeats at high frequencies to create a pitched sound. The timing of the reeds vibration is influenced directly by the length of the tube, which releases pressure based on its own fundamental standing wave.

Reed instruments can have either a single or a double reed. Single reed instruments place the reed against a mouthpiece, which forms the chamber (see Figure 5.4). Double reed instruments have mouthpieces where two reeds are bound together and vibrate against each other.

Reed instruments represent a different type of standing wave, where the pipe is closed at one end. In pipes with just one opening, the wave is restricted

Fig. 5.4: A single-reed mouthpiece. The reed alternately seals and releases from the mouthpiece behind it.

at the closed end. Thus, closed cylinders play one octave below open cylinders of the same length, and have opposite pressures at opposite ends of the tube. If the opening of a clarinet is releasing a high or low pressure wave, the mouthpiece has a neutral pressure state and vice versa (see Figure 5.5).

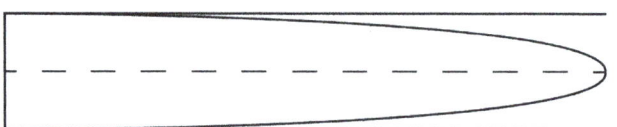

Fig. 5.5: A simple representation of the standing wave of a closed tube.

This opposite quality has an interesting effect on the sounds of closed tubes. Since the pressure is always opposite at opposite ends, the wave is represented by one quarter of a sine wave. Harmonic frequencies of the open tube exist at integer multiples. However, even multiples of the fundamental would not have opposite pressures at the two ends of the tube, harmonic partials of a closed reed instrument only exist at odd multiples of its fundamental frequency (see Figure 5.6).

The clarinet also produces overtones by over-blowing. However, unlike the flute, the clarinet over-blows at the perfect 12th (one octave plus a perfect 5th), which is the third partial, and not the octave.

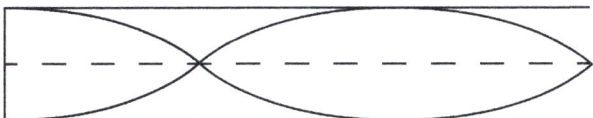

Fig. 5.6: A harmonic frequency of a closed tube. Note that since pressure must be extreme at the closed end, only odd multiples of the fundamental in Figure 5.5 give a harmonic frequency.

5.4 Brass

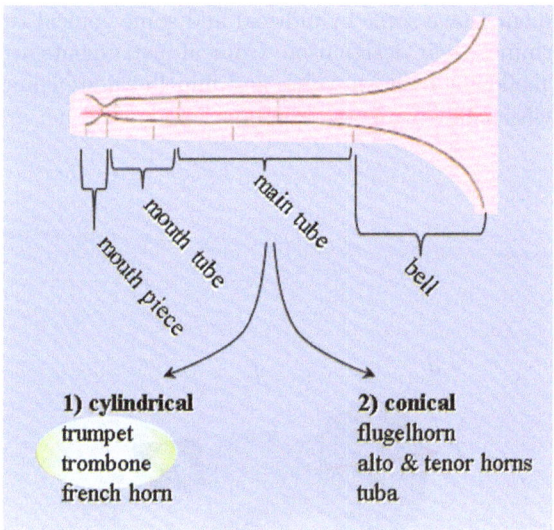

Fig. 5.7: The anatomy of a brass instrument.

Brass instruments are aerophones where the lips of the player function as the oscillator. Using a mouthpiece to create tension, brass players send air through a closed tube which sympathetically vibrates with their lips. As with woodwind instruments the length of the instrument determines the pitch produced. Unlike woodwind instruments, brass instruments do not have tone holes. Multiple pitches are obtained by lengthening the tube or increasing the pressure on the lips, which produces the next partial of the instrument.

Brass instruments, like the clarinet, are closed tubes. This would normally imply that there are specific resonating frequencies that correspond to the odd multiples of the fundamental frequency. This creates a major problem: players would only be able to play pitches that correspond to these limited overtones. Fortunately, there are two major solutions: the mouthpiece and the bell. The mouthpiece lowers the upper partials of the instrument, while the bell raises the lower pitches. Through careful design, these partials are tuned to frequencies equal to $2f, 3f, 4f$, and $5f$, where f is the fundamental frequency of the instrument.

Brass instruments use slides and valves to lengthen the tube. Slides are typically greased sections of tubing that interlock. They can be extended to lower the natural frequency of the tube. The trombone is the most common modern slide instrument. Valves, which are either rotary or piston, redirect the airflow to different sections of tubing that lengthen the distance pressure waves must travel through the instrument. Trumpets, horns, and tubas are all valve instruments. Valves allow for discrete changes in pitch, while slides allow a continuous spectrum of frequency.

Brass instruments are either conical or cylindrical, see Figure 5.7. While all brass instruments have some cylindrical and some conical tubing, the ratio of the two determines their designation. Conical instruments are typically perceived as more mellow. The horn, tuba, and flugelhorn are conical. Cylindrical instruments, such as the trumpet, trombone, and baritone, are brighter and more piercing.

5.5 Strings

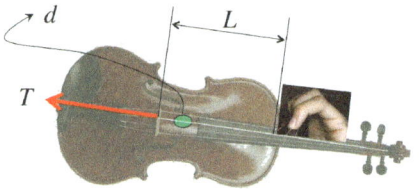

d = density (gr/cm)
L = string length (cm)
T = tension (N)
f = frequency (Hz)

$$f = \frac{1}{2L}\sqrt{\frac{T}{d}}$$

Fig. 5.8: The factors of a violin's sound frequency.

String instruments produce sound by plucking or bowing a string. The pitch of the string is controlled by the string's diameter and tension, see Figure 5.8 for details. A player can adjust the pitch further by holding the string against a fingerboard to shorten the length. The pitch is determined by the placement of the fingers, or by a fret. The new note is determined by the ratio between the length of the shortened string and its original length. A complete discussion of these ratios can be found in the next chapter.

More importantly, the violin family uses resonating cavities to enhance the sound of the vibrating string. The body of any string instrument has specific harmonic frequencies that correspond to the vibrations of the string. They are determined by the shape of the body. These resonant modes, shown in Figure 5.9, enhance the sound and create the timbre difference between different string instruments.

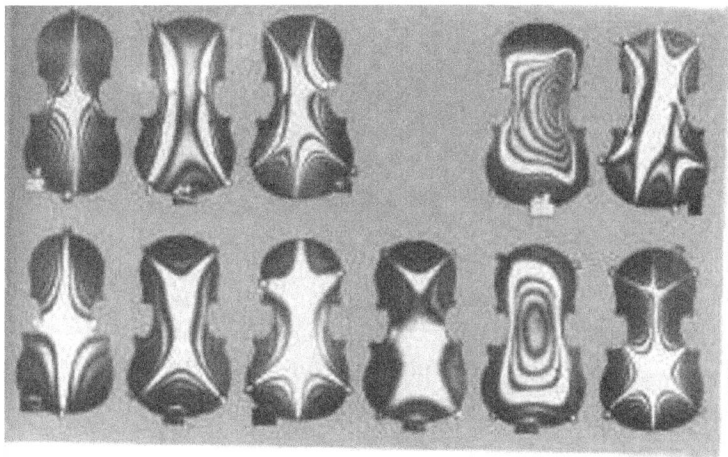

Fig. 5.9: The resonant modes of a violin.

5.6 Percussion

Percussion refers to instruments that are stuck or shaken to produce sound. Depending on the instrument and the performance situation, percussion instruments may be struck with the hands or with sticks, mallets, or beaters specific to the instrument. Striking provides the energy to set up a vibration in the instrument (or triggers an electrical signal) that ultimately produces sound. Percussion is a remarkably diverse group, including instruments that fall into every Hornbostel-Sachs category. For this reason we limit our description to drums (membranophones) and mallet keyboard instruments (idiophones).

Drums are constructed of one or more thin flexible skins (also called *drum heads*) stretched tightly across a stiff frame and/or resonating body. Larger drums with looser membranes generally vibrate at lower frequencies. Some drums are *pitched*, producing periodic sounding waves, which can often be tuned to a desired frequency by tightening or loosening the head. These drums usually consist of a single drum head, or two heads stretched to resonate at the

42   5 Musical Instruments

same frequency. *Unpitched* drums, however, usually have two heads, which are out of tune with each other. The effect is a sound with a vague pitch that fades quickly.

Mallet or keyboard instruments include the xylophone, marimba, vibraphone, and glockenspiel. These instruments produce sound with up to sixty or more vibrating metal or wooden bars. When struck, each bar vibrates at a specific frequency, determined by the bar's length, shape, and material. In typical Western instruments, bars are supported at two points, called nodes, along their length, allowing motion at the ends and in the center. Most mallet instruments are naturally amplified by a set of closed-end tubes, one for each bar. These are tuned to resonate at the same frequency as their respective bar.

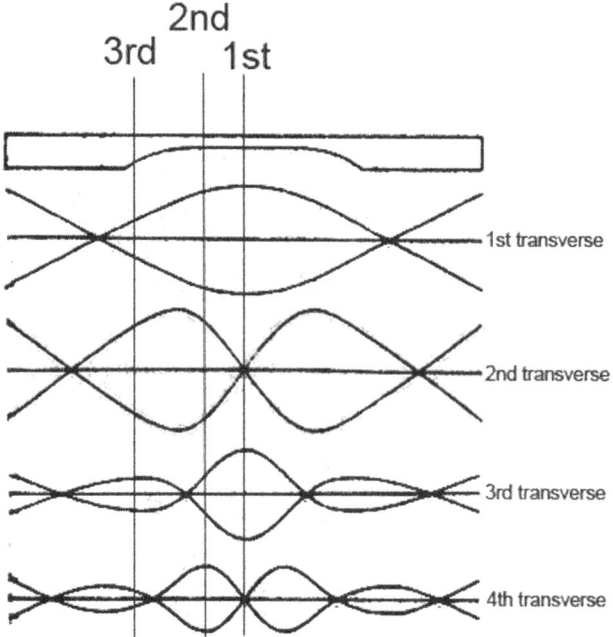

Fig. 5.10: The resonant modes of a marimba bar. The nodes are placed to allow the fundamental to ring freely.

As in other vibrational cases, marimba bars vibrate in several modes (see Figure 5.10) simultaneously to create a rich and pleasant set of overtones. The frequency of all the overtones can be raised by sanding the ends of a bar to make it shorter. Removing a small amount of material can noticeably change the pitch. Scooping out material in the middle of the bar makes it more flexible, effectively lowering the pitch. Since the fundamental mode requires the most motion in the center, removing material here tunes the fundamental. Material can also be removed between the middle and the nodes to tune the higher overtones. High-end marimba makers typically give attention to at least the first three overtones, resulting in so-called "triple-tuned" instruments.

5.7 Piano

The piano is known as a particularly versatile instrument, prominent in both accompanied and unaccompanied performance in many diverse genres. Its sound originates in vibrating strings; however, these vibrations are made audible by a large vibrating soundboard. In the Hornbostel-Sachs system, the piano may thus be considered an idiophone, a chordophone, or both. When a piano key is pressed, a somewhat complicated mechanical system translates this press into the motion of a small felted hammer. The hammer then strikes a string or group of strings, establishing a vibration. Typical pianos have eighty-eight such hammers, one for each key or distinct pitch. The piano is designed to be sensitive to variations in touch, so a harder key-press makes the hammer move faster. This allows for variations in the loudness of the instrument (hence its full name, the *piano forte*).

Fig. 5.11: The piano's anatomy: (1) frame, (2) lid, front part, (3) capo bar, (4) damper, (5) lid, back part, (6) damper mechanism, (7) sostenuto rail, (8) pedal mechanism, rods, (9,10,11) pedals: right (sustain/damper), middle (sostenuto), left (soft/unacorda), (12) bridge, (13) hitch pin, (14) frame, (15) soundboard, (16) string.

The inner structure of the piano (see Figure 5.11) is just as critical as the keys and hammers. The strings, which are typically metal, are attached to the instrument by a strong metal frame. This frame must sustain a tremendous amount of tension, typically around seven tons, as each string must be stretched tightly to be in tune and produce a good tone. The thickness of each string also affects its pitch. Since these metal strings do not have much surface area, they cannot create large enough air pressure waves to be audible in typical performance situations on their own. The vibration of the string is therefore transferred via a supporting piece called the *bridge* to the soundboard. This is a wooden structure made of several layers and tuned to resonate at specific desirable frequencies. The soundboard typically takes up a majority of the space in the piano, giving it plenty of surface area.

The piano also has pedals, which influence how the hammers interact with the strings. The leftmost pedal lowers the *action* of the hammer, bringing the head closer to the string. This allows a player to play more quickly, since less time is required for the hammer to travel from its original position. The center pedal controls sustain; pressing it removes damping from the string, allowing it to ring freely. The rightmost pedal affects the volume of the piano. In many pianos, pressing this pedal simply lowers a piece of fabric between the hammers and the strings. This dampens the sound of the instrument.

5.8 Voice

The human voice is a remarkable and complex instrument, consisting of many mechanisms in the chest, throat, and head that work together to produce a rich, distinct sound. Several well-defined methods of training the body to produce an ideal singing voice exist. However, the purpose of this section is simply to describe the various components that contribute to the human voice.

Sound production begins with inhaling air into the lungs. To this end, the diaphragm muscle, located between the lungs and the lower abdomen, contracts to expand the lungs. Other muscles in the chest then contract to control the rate of exhalation.

Air passes from the lungs up through the esophagus to the larynx, which is the origin of the vibrations for both singing and speaking. The exact shape of the larynx is unique to every individual, giving everyone slightly different sound characteristics. Structures within the larynx include the epiglottis and vocal folds (see also Figure 5.12). The epiglottis is a cartilage flap that prevents things besides air from entering the lungs. The vocal folds are flaps of tissue that can be stretched in various ways to produce vibrations when air passes through them. They can be opened completely while breathing, or closed tightly when producing strained or grunting sounds. This structure along with the breath is primarily responsible for the pitch and loudness of a sung note.

The vocal folds vibrate in a variety of modes, commonly referred to as *registers*. They have been classified according to different methods of the vocal community, identifying anywhere between one and six distinct vibrational modes. Registers are differentiated by the thickness and tension of the vocal folds, and sometimes by which additional *resonating chambers* are also in use. One common theory, the two-register theory, distinguishes *chest voice* from *head voice*. Chest voice is perceived by the singer as resonance in the lower throat and/or chest, and results from vibration of the majority of the vocal folds. Chest voice is generally considered to sound fuller. In contrast, head voice uses

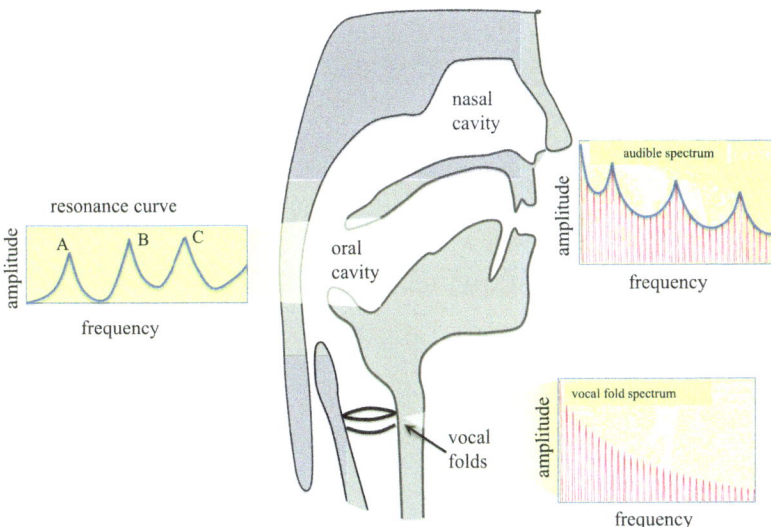

Fig. 5.12: The human voice anatomy and its spectral shaping. The vocal folds generate a uniform spectrum, which is filtered in the oral and nasal cavities. To the left, we see the three formants A, B, and C.

a smaller percentage of the folds and is perceived by the singer as resonance in the upper throat and/or head. The latter is generally considered to sound softer or lighter.

Sound is amplified as it passes through the pharynx, which connects the mouth to the larynx. Sound resonates here and in the sinus cavities to produce a distinctly human voice. Proper training and use of these cavities results in good vocal tone and projection. Finally, articulations, such as consonant sounds, are created using the tongue, teeth, lips, and cheeks. The frequency distribution of the human voice is described by three peaks of spectral amplitudes, the formants A, B, and C (see left part of Figure 5.12).

The relative intensity of formants A and B determines the pronunciation of words (Figure 5.13). Singers have special high (C) formants ($2,500\ Hz$), which enable them to be heard even with loud orchestras. When the fundamental frequency of a pitched sound is higher than the first formant (A) frequency, the singer may change the formant to a higher value to keep the fundamental in audible loudness.

5.9 Electronic Instruments

5.9.1 Theremin

The earliest electronic musical instrument is called the *theremin* (originally *aetherophone*). It was invented by Leon Theremin in 1920, see Figure 5.14,

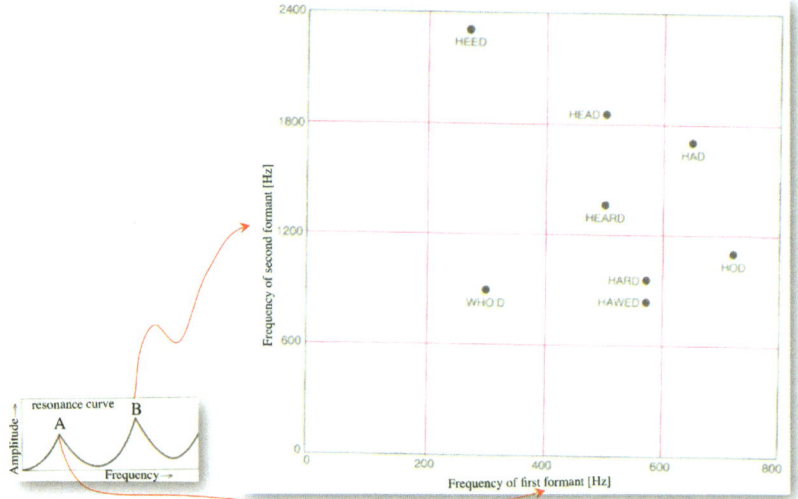

Fig. 5.13: The pronunciation of similar words is controlled by a specific distribution of the relative intensities of formants A and B.

Fig. 5.14: Leon Theremin.

sound example: `theremin.mp3`. It is the only musical instrument that can be played without being touched by the performer. A theremin consists of two oscillators with one plate each. One oscillator runs at a fixed frequency and the other runs at a variable frequency pretty close to the fixed one. The hands and body of the musician form the second plate of the oscillators. Therefore, by moving one's hands closer to or further away from the fixed plates, they change the capacity of the oscillator and consequently alter the frequency of the produced sound. The instrument is responsive only to movements that change the distance between the hands and the oscillator plates. Any different motion has no influence on the produced sound. The theremin is considered the predecessor of modern synthesizers, since it was the first electronic instrument to offer the performer complete control of two critical concepts of music: pitch and volume.

5.9.2 Trautonium

Fig. 5.15: The Trautonium.

The Trautonium is an electronic instrument that was invented by Friedrich Trautwein in 1929, in Berlin (see Figure 5.15). A Trautonium produces sound via the touch of a resistor wire on a metal plate, sound example: `trautonium.mp3`. Pitch depends on the position of the contact point on the string whereas loudness is determined by the amount of pressure that is used by the performer as well as by a foot pedal. The soundwaves produced by the Trautonium are saw-like. The instrument offers extensive playing techniques such as vibrato or portamento, which depend on the motion of the fingers of the performer. The most famous Trautonium piece is the soundtrack of *The Birds* by Alfred Hitchcock.

5.9.3 U.P.I.C.

Unité Polyagogique Informatique CEMAMu is a computerized music composition interface that was conceived by composer Iannis Xenakis during the 1950s in Paris (see Figure 5.16). Xenakis was composing his famous piece *Metastasis*, in which he used graphics to describe sound effects that were impossible to describe with music notation. The arduousness of writing 61 different graphic parts for the entire orchestra, as well as translating them into conventional notes for the orchestra to be able to perform them, made him come up with the idea of a computer system that could draw music. U.P.I.C. was developed

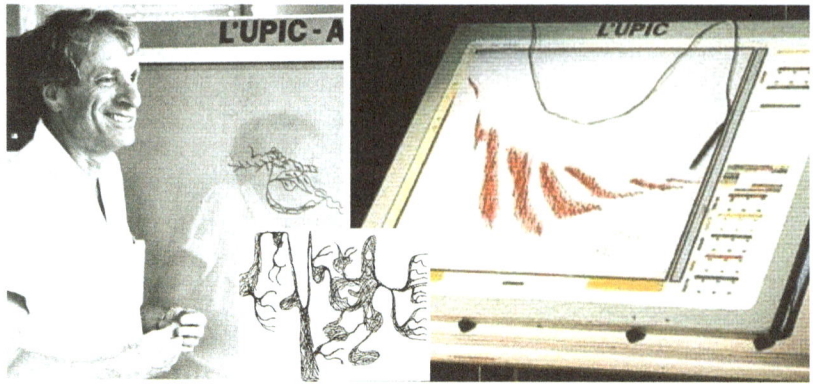

Fig. 5.16: Xenakis with his U.P.I.C.

at Centre d'Etudes de Mathématique et Automatique Musicales (CEMAMu) in Paris. U.P.I.C. consists of an electronic tablet with a vectorial display. The tablet is connected to a computer that renders the graphic input. Basically, the performer creates and stores waveforms that can later be used in real time for the production of sound. U.P.I.C. offers extremely deep and detailed manipulation of sound.

5.9.4 Telharmonium or Dynamophone

Fig. 5.17: The Telharmonium.

The Telharmonium is an electric organ created by Thaddeus Cahill in 1896 in New York (see Figures 5.17 and 5.18). It is considered to be the ancestor of modern synthesizers. Its mechanism included the generation of electrical oscillations identical to the acoustic vibrations of the desired tones. In other words, the instrument was a huge power generator that made electrons vibrate with the same sinusoidal function as the vibration of air molecules of a musical tone.

Fig. 5.18: Details of the Telharmonium.

To make them audible, these electrical signals were transmitted through wires in the same way as signals through telephone circuits. In addition, the instrument had the ability to form sounds in detail, especially woodwind instrument sounds such as the flute and clarinet.

Eventually, the Telharmonium's popularity declined for a number of reasons that included its huge size, weight (the instrument weighed more than 200 tons and occupied an entire room), and power consumption. In many cases, telephone broadcasts of the instrument would interfere with other broadcasts, and telephone users would have their call interrupted by bizarre electronic music.

5.9.5 MUTABOR

Fig. 5.19: The MUTABOR; Rudolf Wille is third from left.

MUTABOR is software developed in 1980 at the Technical University of Darmstadt under the direction of mathematician Rudolf Wille (see Figure 5.19). MUTABOR was designed to alter the tuning of musical instruments with microtonal precision. It enables the construction of microtonal scales and different types of tuning that are far from the Western music tradition. Moreover, MUTABOR offers musicians the opportunity to experiment with intervals uncommon in Western music and provides a mathematical language for contemporary music [47].

6
The Euler Space

Summary. In this chapter, we will discuss music theories from two mathematical perspectives. In the first part, we will explain what is a tuning and how the Euler Space relates to Western tuning systems. We will use Euler's theory of tuning to explain three typical tunings: *equal temperament* tuning, *Pythagorean* tuning, and *just* tuning. In the second part, we will use a mathematical torus model $\mathbb{Z}_3 \times \mathbb{Z}_4$ and its symmetries to explain counterpoint and its application in the RUBATO® music software.

$$- \Sigma -$$

6.1 Tuning

6.1.1 An Introduction to Euler Space and Tuning

The Euler Space is one of the most important contributions to music theory by Leonhard Euler, a Swiss mathematician, physicist, astronomer, logician, and engineer (see Figure 6.1) [17]. It defines the geometric space of pitch classes, where a two-dimensional space is spanned by the axis of fifths and the axis of major thirds (see Figure 6.2).

Tuning, on the other hand, is defined as a process of frequency adjustment to pitch. Different tuning systems use different methods to define frequencies of given pitches, but they always start with a set pitch value (frequency), typically A4 to be 440 *Hz*. Each tuning system has its advantages and disadvantages. The most common system is called *twelve- (or equal-) tempered tuning*. This is the tuning system that people currently use on pianos; it divides an octave into twelve equal semitones, and that is where the name comes from. Equal temperament was first introduced by the Chinese mathematician Zhu Zaiyu in 1584 (its precise definition is given in Section 6.1.2.1).

In this way, 12-tempered tuning eliminated the problem of changing tonality in musical modulation. Equal temperament is especially important to instruments such as the piano, because it allows composers to explore different

Fig. 6.1: Leonhard Euler (1707-1783), who was a Swiss mathematician, physicist, astronomer, logician, and engineer, and made many crucial discoveries in mathematics.

keys freely. However, at the same time, one also loses the flavor and color of some intervals in Pythagorean and just tuning (see Sections 6.1.2.2 and 6.1.2.3).

Of course, lots of other tuning approaches also exist. For example, meantone tuning, well-tempered tuning, and mircotonal tuning.

6.1.2 Euler's Theory of Tuning

The general relationship between pitch and frequency f is given by the formula

$$pitch(f) = log(f_0) + o \cdot log(2) + q \cdot log(3) + t \cdot log(5) \sim o \cdot log(2) + q \cdot log(3) + t \cdot log(5),$$

where the coefficients o, q, and t are rational numbers, and f_0 is a reference frequency. Since these numbers are unique for a given pitch, we can think of them as being coordinates in a three-dimensional space. This is the *Euler Space of pitches*. The three axes represent octave, fifth, and major third, respectively.

Euler was the first to present music pitches in a geometric way. He first mentioned this geometric representation in a drawing (see Figure 6.2), where the horizontal lines represent intervals of fifths, and the vertical lines represent major thirds.

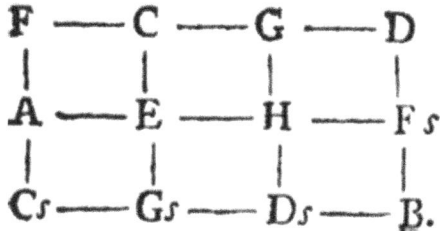

Fig. 6.2: The Euler Space of fifths (horizontal axis) and major thirds (vertical axis).

6.1.2.1 Equal Temperament

For this tuning, the coordinates q and t vanish, and we only have the coefficient $o = n/12$, where n is an integer, see Figure 6.3. The column to the right shows the pitch in Cents (Ct). A Cent is one hundredth of a semitone.

Tone name	Frequency ratio	Octave coord.	Fifth coord.	Third coord.	Pitch (Ct)
c	1	0	0	0	0
$d_\flat$	$2^{1/12}$	1/12	0	0	100
d	$2^{2/12}$	2/12	0	0	200
$e_\flat$	$2^{3/12}$	3/12	0	0	300
e	$2^{4/12}$	4/12	0	0	400
f	$2^{5/12}$	5/12	0	0	500
$f_\sharp$	$2^{6/12} = \sqrt{2}$	6/12	0	0	600
g	$2^{7/12}$	7/12	0	0	700
$a_\flat$	$2^{8/12}$	8/12	0	0	800
a	$2^{9/12}$	9/12	0	0	900
$b_\flat$	$2^{10/12}$	10/12	0	0	1000
b	$2^{11/12}$	11/12	0	0	1100

Fig. 6.3: The frequency ratios for equal temperament.

6.1.2.2 Pythagorean Tuning

Back at the beginning of music-tuning history, *Pythagorean tuning* was first discovered around the sixth century B.C. by Pythagoras. Pythagorean tuning is a tuning system based on the ratios of the first four numbers 1, 2, 3, 4. These numbers are shown in the Pythagorean tetractys (see Figure 6.4). We

54   6 The Euler Space

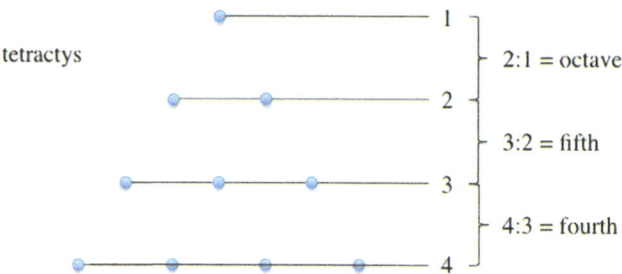

Fig. 6.4: The Pythagorean tetractys.

see that the ratios of this configuration define important intervals: octave, fifth, and fourth. To define the chromatic scale in this tuning, we are only allowed to take powers of 2 and 3. Figure 6.5 shows the ratios for the chromatic scale tones. They can be very complicated, but this tuning dominated European music until the Middle Ages.

Tone name	Frequency ratio	Octave coord.	Fifth coord.	Third coord.	Pitch (Ct)
c	1	0	0	0	0
$d_\flat$	256/243	8	-5	0	90.225
d	9/8	-3	2	0	203.91
$e_\flat$	32/27	5	-3	0	294.135
e	81/64	-6	4	0	407.82
f	4/3	2	-1	0	498.045
$f_\sharp$	729/512	-9	6	0	611.73
g	3/2	-1	1	0	701.955
$a_\flat$	128/81	7	-4	0	792.18
a	27/16	-4	3	0	905.865
$b_\flat$	16/9	4	-2	0	996.09
b	243/128	-7	5	0	1109.78

Fig. 6.5: The frequency ratios for Pythagorean tuning.

6.1.2.3 Just Tuning

Just (or pure) tuning is based on powers of 2, 3, and 5. The last number 5 was introduced in the Middle Ages to get a better major third, namely with

Tone name	Frequency ratio	Octave coord.	Fifth coord.	Third coord.	Pitch (Ct)	% deviation
c	1	0	0	0	0	0
$d_\flat$	16/15	4	-1	-1	111.73	+11.73
d	9/8	-3	2	0	203.91	+1.96
$e_\flat$	6/5	1	1	-1	315.65	+5.22
e	5/4	-2	0	1	386.31	-3.42
f	4/3	2	-1	0	498.05	-0.39
$f_\sharp$	45/32	-5	2	1	590.22	-1.63
g	3/2	-1	1	0	701.96	+0.28
$a_\flat$	8/5	3	0	-1	813.69	+1.71
a	5/3	0	-1	1	884.36	-1.74
$b_\flat$	16/9	4	-2	0	996.09	-0.39
b	15/8	-3	1	1	1088.27	-1.07

Fig. 6.6: The frequency ratios for just tuning.

frequency ratio 5:4. Figure 6.6 shows the representation of the chromatic scale in just tuning. Figure 6.7 shows the chromatic scale as a point set in the Euler space.

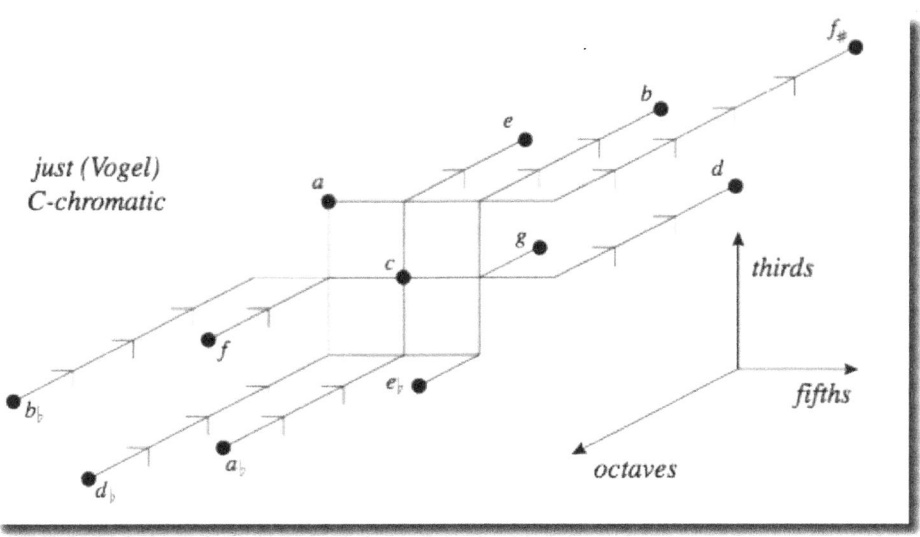

Fig. 6.7: The chromatic scale over c as a point set in the Euler Space.

6.2 Contrapuntal Symmetries

In classical counterpoint following Johann Joseph Fux [21], several fundamental facts are not understood properly. The first fact is that the consonant intervals include the following types (numbers of semitones in parentheses): minor third (3), major third (4), fifth (7), minor sixth (8), major sixth (9), and perfect eighth (12) (or unison (0)). The remaining six dissonant intervals are the minor second (1), major second (2), fourth (5), tritone (6), minor seventh (10), and major seventh (11). The problem is that the fourth is classically viewed as being consonant. This relates to the Pythagorean approach, later extended to Zarlino's harmony, which in terms of Fourier partials reads as follows. The fundamental frequency f and its multiples $2f, 3f, 4f, 5f$ define consonant intervals by fractions of successive partials: $2f : f$ is the octave, $3f : 2f$ is the fifth, $4f : 3f$ the fourth, and $5f : 4f$ the major third. This is also realized for the just tuning. This principle contradicts the dissonant role of the fourth. The second problem of counterpoint is the rule of forbidden parallel fifths. The argument is psychological rather than theoretical, and why are parallel thirds not forbidden?

In order to solve these problems, we propose a different geometric representation of pitch classes and intervals using a torus (Figure 6.8). This geometry will allow a purely mathematical theory of counterpoint that also solves the two problems mentioned above. Moreover, this approach has also been implemented in a counterpoint component of composition software RUBATO®, see also Chapter 16.

6.2.1 The Third Torus

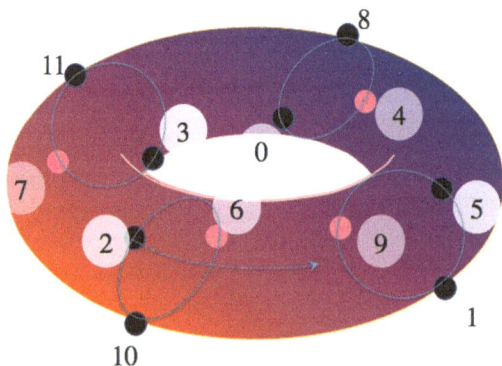

Fig. 6.8: The third torus. We see four vertical circles—copies of $\mathbb{Z}_3$—attached to the four pitch-class points $0, 3, 6, 9$.

6.2 Contrapuntal Symmetries

The third torus, as shown in Figure 6.8, has two circles, one, $\mathbb{Z}_3$, with three values, the other, $\mathbb{Z}_4$, with four values. Every pitch class $0, 1, 2 \ldots 11$ in the pitch-class set $\mathbb{Z}_{12}$ (the clock face), can be represented by a unique point on the torus as follows. Write a pitch class x as a sum of x_3 major thirds (4) and x_4 minor thirds (3). Can every pitch class be written this way? To begin with, we have $1 = 13 = 1 \cdot 4 + 3 \cdot 3$, which holds in the pitch-class circle $\mathbb{Z}_{12}$. Therefore every multiple of 1 is also such a sum. The point of the torus corresponding to x is the pair (x_3, x_4) of these third multiple numbers as represented by the points on the two circle types $\mathbb{Z}_3, \mathbb{Z}_4$. The torus, which consists of all these pairs of numbers, is therefore denoted by

$$T_{3\times 4} = \mathbb{Z}_3 \times \mathbb{Z}_4$$

In this expression, $\mathbb{Z}_4$ means dividing an octave into three groups of four pitch classes, each three semitones (minor third) apart: $\{0, 3, 6, 9\}, \{1, 4, 7, 10\}, \{2, 5, 8, 11\}$. Then we can establish a base circle by four pitch-class points 0,3,6,9. Similarly, from $\mathbb{Z}_3$ we divide an octave every four semitones (major third), see Figure 6.9. By vertically attaching $\mathbb{Z}_3$ circles to 0,3,6,9, we can establish the torus model with all 12 pitches (Figure 6.8).

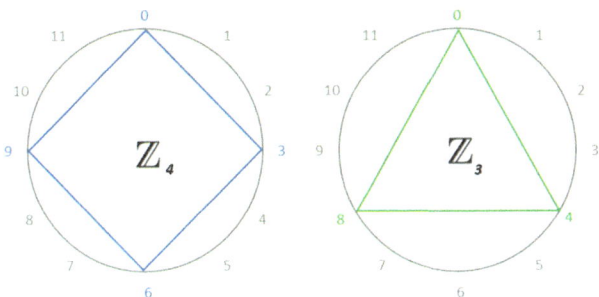

Fig. 6.9: $\mathbb{Z}_4$ means dividing an octave into three groups of four pitch classes, each three semitones (minor third) apart. Then we can establish a base circle by four pitch-class points 0,3,6,9. Similarly, from $\mathbb{Z}_3$ we divide an octave every four semitones (major third).

On the third torus $T_{3\times 4}$, we can define a metrical distance by $d(z,w)$ between points z and w, which is the minimal number of minor or major third steps to reach w from z. The third steps are addition or subtraction of $(1_3, 0_4)$ or $(0_3, 1_4)$. For example, according to Figure 6.8, if we want to go from 3 to 8, the path is 3 to 0 to 8, which goes through one part of a $\mathbb{Z}_3$ path and one part of a $\mathbb{Z}_4$ path (for example, there is ± a minor third and ± a major third between D♯ and G♯), which is also represented as $-1 \cdot (1_3, 0_4) - 1 \cdot (0_3, 1_4)$. The distance from 3 to 8 is calculated by adding up the absolute values of the coefficients of $(1_3, 0_4)$ and $(0_3, 1_4)$. So $d(3, 8) = 2$. As a result, this three-dimensional model can be applied for calculating intervals by using major or minor thirds.

This metrical distance geometry is important because it is used for explaining symmetries of $T_{3\times 4}$. Let us now explain what symmetries of $T_{3\times 4}$ are. On $\mathbb{Z}$ we considered symmetries of the form $T_{\pm 1}^t(z) = t + (\pm 1)z$. This means that we considered the invertible elements ± 1 (-1 means inversion) of $\mathbb{Z}$ and used multiplication with such invertible elements, together with a transposition by t.

When it comes to the torus, not only ± 1 but four elements are invertible, namely $1, 5, 7, 11$. In fact, we have $1^2 = 5^2 = 7^2 = 11^2 = 1$.[1] All other elements are not invertible, for example $3 \cdot 4 = 12 \to 0$, $4^2 = 16 \to 4$. With this multiplicative structure we now generalize the above construction of symmetries. We define symmetries of $\mathbb{Z}_{12}$ by the expression $T_s^t, t \in \mathbb{Z}_{12}, s \in \{1, 5, 7, 11\}$, setting $T_s^t(z) = t + sz$ for $z \in \mathbb{Z}_{12}$.

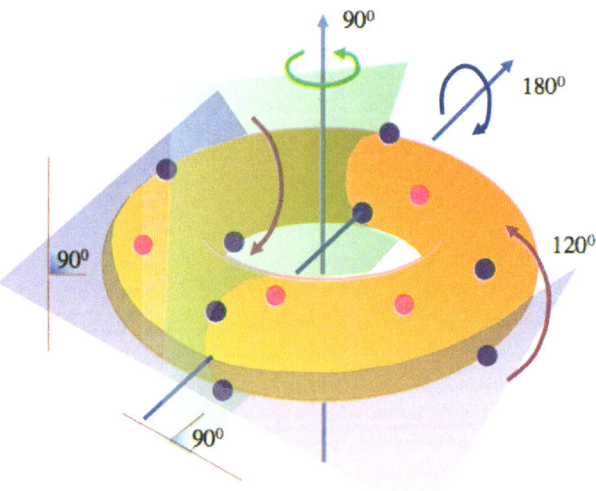

Fig. 6.10: The symmetries of the torus all conserve the third distances.

Figure 6.10 geometrically proves the invertibility of the above symmetries when represented as actions on the third torus. The transposition T^3 translates into a rotation of 90° around the vertical middle axis of the torus. The transposition T^4 translates to a 120° tilting movement of the torus. An inversion T_{11}^0 becomes a 180° rotation of the torus around the horizontal axis through

---

[1] We say these numbers are equal based on their remainders when dividing by 12: $1^2 = 1 \to 1, 5^2 = 25 = 2 \times 12 + 1 \to 1, 7^2 = 49 = 4 \times 12 + 1 \to 1, 11^2 = 121 = 10 \times 12 + 1 \to 1$.

0 and 6. The symmetry T_5^0 becomes a reflection of the torus in the horizontal plane through the middle of the torus. The symmetry T_7^0 becomes a reflection of the torus in the vertical plane through 0 and 6. This means that all symmetries are combinations of such classical torus symmetries (such as reflection and rotation).

This geometric music model was introduced in [33]. The biggest advantage of this model is that the third distances $d(z, w)$ are conserved under all the above torus symmetries. For example, if we take notes C and G (0 and 7). $d(C, G) = d(0, 7) = 2$. If the torus rotates 90° around the vertical middle axis by transposition T^3, then we get $0 + 3 = 3$ and $7 + 3 = 10$. We still get $d(3, 10) = d(D\sharp, A\sharp) = 2$. If the torus is reflected in the horizontal plane through 0 and 6 with symmetry T_5^0, we get $0 \times 5 = 0$ and $7 \times 5 = 35 \to 11$. $d(0, 11) = d(C, B) = 2$.

6.2.2 Counterpoint

In counterpoint, one starts with the construction of a composition from two voices: cantus firmus (CF) and discantus (D). The rules were first developed in ninth century Europe, and were quite established in the sixteenth century. There are five so-called *counterpoint species* with increasing complexity, but here we only focus on the most basic first one. For each CF note the first species only admits one D note a consonant interval apart from the CF note, see Figure 6.11 for an example.

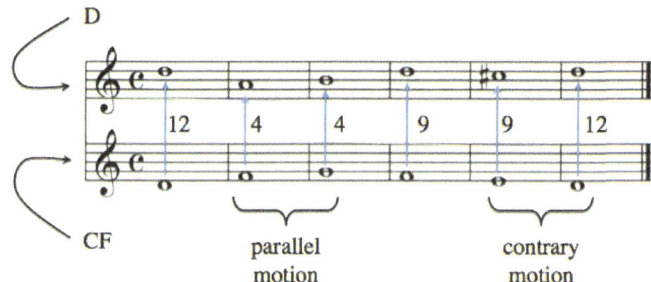

Fig. 6.11: An example of a first-species counterpoint.

Using this geometric representation on the torus, we can now solve the problem of the dissonant fourth mentioned above. The distribution $K = \{0, 3, 4, 7, 8, 9\}, D = \{1, 2, 5, 6, 10, 11\}$ (also known as the Fux dichotomy (K, D)) of intervals in consonant and dissonant halves can be constructed without an invalid reference to acoustical arguments. Instead, we now define consonances and dissonances as members of interval *sets*, and not as properties of individual intervals.

6 The Euler Space

We find a solution by observing the symmetry model we established above. There is a *unique* symmetry $AC = T_5^2$ of $\mathbb{Z}_{12}$ that transforms K into D. More precisely,

$$AC(0) = 5 \cdot 0 + 2 = 2, AC(3) = 5 \cdot 3 + 2 = 5, AC(4) = 5 \cdot 4 + 2 = 10,$$
$$AC(7) = 5 \cdot 7 + 2 = 1, AC(8) = 5 \cdot 8 + 2 = 6, AC(9) = 5 \cdot 9 + 2 = 11$$

and vice versa $AC(D) = K$ since AC^2 leaves every number fixed. For example, $AC^2(5) = AC(5 \cdot 5 + 2) = AC(3) = 5 \cdot 3 + 2 = 5$. This is the *autocomplementarity function*. There are five other interval dichotomies (X, Y) that have *unique* autocomplementarity functions; we call them *strong dichotomies*. But why has the Fux dichotomy been commonly used for centuries? Let us see the distribution of the Fux dichotomy on the third torus (Figure 6.12). From this result we can learn that elements in set K are greatly separated from elements in set D in the Fux dichotomy, whereas for the other five strong dichotomies of K and D, elements are mixed up on the third torus. This distinguishes the Fux dichotomy from others.

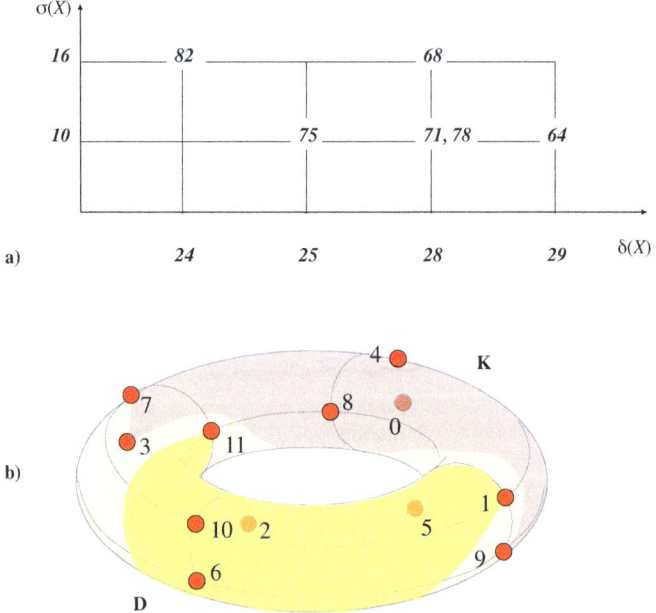

Fig. 6.12: The diameters (average distance between points of consonances) and spans (average distances between consonances and their associated dissonances via the autocomplementarity function) of the six strong dichotomies are shown in (a), and the geometric configuration for the Fux dichotomy (K, D) is shown in (b).

By complex computations with torus theory, we can also explain why parallel fifths are forbidden, and why parallel thirds are allowed. Further explanation can be found in [37, Part VII].

These symmetries are implemented in the music software RUBATO®. In RUBATO®, users assign tasks to each Rubato component (rubette), then connect rubettes to build a RUBATO® network [3]. RUBATO® can generate counterpoint compositions based on the corresponding counterpoint rules. All the six strong dichotomies we mentioned above are implemented in the counterpoint rubette. Users can choose different dichotomies to write counterpoint in unconventional ways. See Chapter 16 for more information about RUBATO®.

Part III

Electromagnetic Encoding of Music: Hard- and Software

7
Analog and Digital Sound Encoding

Summary. This chapter gives an overview of the most prominent analog and digital sound encoding systems of the present music industry.

$$- \Sigma -$$

7.1 General Picture of Analog/Digital Sound Encoding

The mechanical and electromagnetic encoding of music is the trace of an incredibly rich history of creative acts. We want to follow this pathway and highlight the most important milestones.

The general scheme of such encoding is shown in Figure 7.1. The scheme shows that this encoding technology inhabits a physical reality and mainly realizes a movement from the poietic original code given by the music's sound waves to a neutral niveau (level) of communicative transfer, where the message is encoded in either analog or digital fashion. "Analog" means that the original waveform is mapped onto a similar waveform imprinted on a medium, and "digital" means that the original waveform is transformed into a sequence of numbers according to a more or less refined algorithm. The neutral niveau is in the expressive layer of the semiotic system; it has no meaning as such, but needs to be decoded in order to give us back the original musical sound content. Observe that "content" here means just the sounding wave, not the deeper symbolic or psychological sound signification. Technically speaking, we have the arrangement shown in Figure 7.2. The physical content is recorded by a set of microphones (short: mics), and sent to an analog mixer, which merges the mics. The mix is then piped to an analog or digital encoder, which then—after potentially mixing the digital material—outputs its analog decoded wave to a second analog mixer for voltage output to a system of loudspeakers.

Although the microphone and loudspeaker technology looks marginal in the music encoding process, it has been a major concern to obtain optimal input from or output to the acoustic level. Figure 7.3 shows the three major mic types:

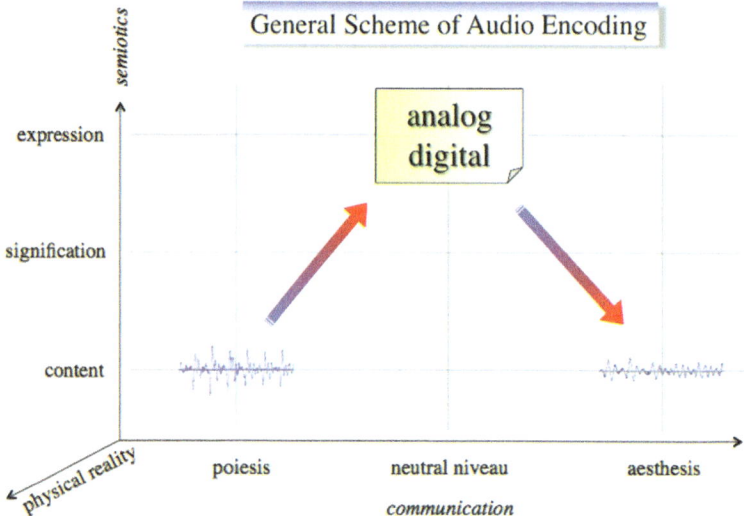

Fig. 7.1: The general scheme of mechanical and electromagnetic encoding of music. The encoding technology inhabits a physical reality and mainly realizes a movement from the poietic original code given by the music's sound waves to a neutral niveau level of communicative transfer, where the message is encoded in either analog or digital fashion.

a) moving coil dynamic, b) ribbon, and c) condenser. While the technologies defining these devices are evident, their use in music recording studios or for concert amplification is subject to sophisticated discussions that also reveal personal tastes and 'ideological' positions concerning the characteristic quality of sound. It is remarkable how much effort has been invested in creating better loudspeaker systems or better mics and marketing such systems under the title of creative technologies. Figure 7.4 shows the typical anatomy of a loudspeaker device.

The concept of a microphone and loudspeaker as technological surrogates for ears and instruments has a more creative implication for the technology of sound recording and reproduction, however. Why should we have only one recording device, or two as in the human ear configuration? The answer is that human ears are also just one (pair), so humans do not need more than a corresponding mic (pair) to record music. This paradigm of high fidelity is also used in audio label advertising (e.g., the excellent Cadence Jazz Records): "We guarantee the most faithful real music recordings."

But where do you listen when hearing a musical performance? Are you simulating a concert hall position, and where in such a hall? The question alloiws a variety of answers: One can sit in an ordinary seat, also creating a variable listening experience according to where the seat is located, and in which hall one wants to be seated. But the variety of answers is much larger in

7.1 General Picture of Analog/Digital Sound Encoding

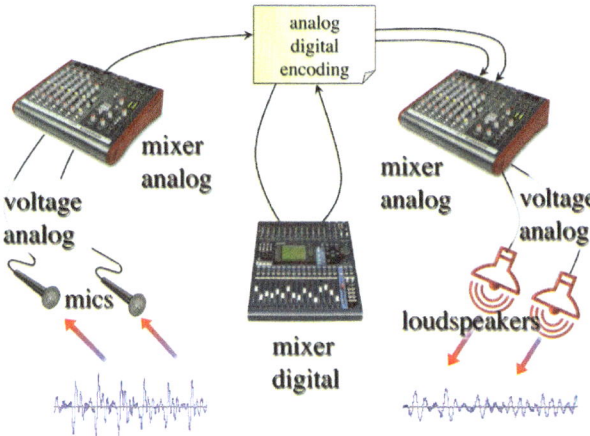

Fig. 7.2: The physical content is recorded by a set of microphones (short: mics) and sent to an analog mixer that merges the mics. The mix is then piped to a analog or digital encoder, which then—after potentially mixing on the digital material—outputs its analog decoded wave to a second analog mixer for voltage output to a system of loudspeakers.

that it is also possible to listen to the music as a musician, really near to some instrumental configuration (within the string choir, for example, or next to a drum, etc.). Which is the 'real' listening position? In studio technology, this is solved by a heavy intermediate process involving the installation of a large number of mics across the orchestra, recording all these tracks and then mixing them while stepping to postproduction. The multiplicity of such artificial ears is even potentiated by positioning some twelve mics across the drum set, for example.

Ideally one would wish to have mics everywhere in the orchestra's space. Of course, this would presuppose a new mic technology that is capable of tracing an infinity of mic locations and merging them into a postproductive synthesis.

But it is clear that this opening of the hearing concept redefines musical reality altogether: It is no longer evident what is the audio output. The variety of local hearing inputs enables a very creative reconstruction of the typical final stereo output.

On the other hand, thinking of the concept of a loudspeaker as a surrogate of instrumental output is also analogous to thinking of the two ears as two sources, or four in quadro setups. But similarly to the theoretically infinite multiplicity of mics, there is also the option of an infinity of loudspeakers distributed in any manifold within the hall or living room of audio reproduction. This infinity might be realized as a surface mic, i.e., a surface that vibrates in a locally variable way. A technology using an infinity of hearing positions might be realized in headphones whose output depends on their position in space and

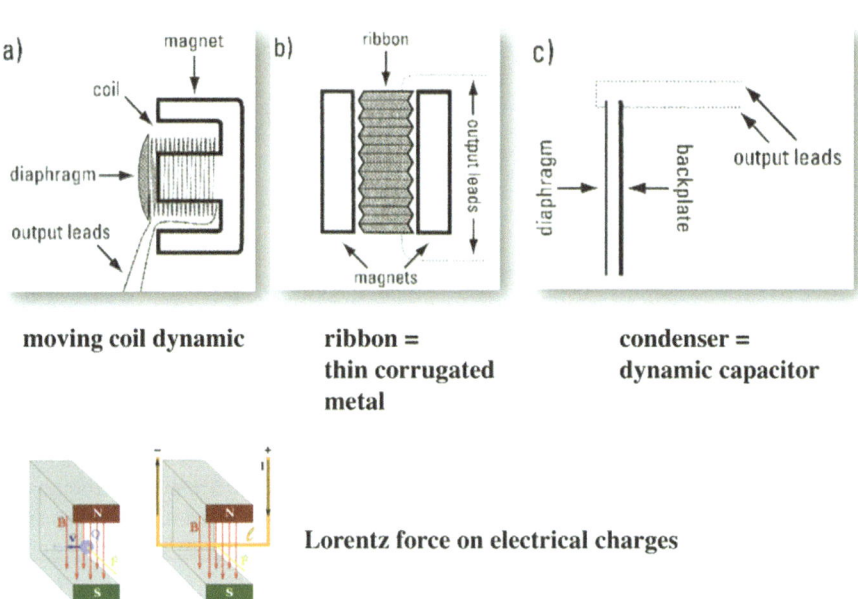

Fig. 7.3: Three major mic types: moving coil dynamic, ribbon, and condenser.

a Bluetooth wireless input of sounds that have been recorded and/or edited to represent spatial positions in the orchestra's recording space.

The overall analog and digital encoding variants are shown in Figure 7.5. For analog encoding, we have the options of a mechanical LP and an analog magnetic tape. For digital encoding, we have three options: (1) the optical technology of a CD, where the data is engraved in a glass material (see below for the details); (2) the magnetic technology realized by DAT (Digital Audio Tape) or Hard Drive technology; (3) the electronic (semiconductor-based) technology with volatile (RAM, Random Access Memory), semi-volatile (memory sticks), and non-volatile (ROM, Read-Only Memory) memory.

7.2 LP and Tape Technologies

Figure 7.6 shows the classical analog encoding and its historical highlights. To graphically represent sound on discs of paper, a machine using a vibrating pen (without the idea of playing it back) was built by Édouard-Léon Scott de Martinville in 1857. While this *phonautograph* was intended solely to visualize sound, it was recently realized that this depiction could be digitally analyzed and reconstructed as an audible recording. An early phonautograph, made in 1860, and now the earliest known audio recording, has been reproduced using computer technology.

7.2 LP and Tape Technologies

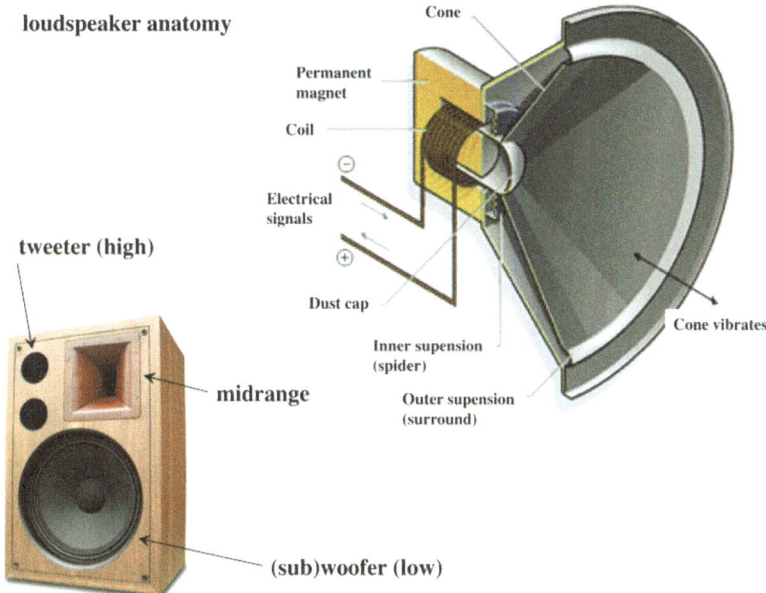

Fig. 7.4: Typical anatomy of a loudspeaker device.

In 1877, Thomas Edison developed the *phonograph*, which was capable of both recording and reproducing the recorded sound. There is no evidence that Edison's phonograph was based on Scott de Martinville's phonautograph. Edison originally recorded on wax-coated tape while Scott de Martinville used soot-coated glass. The version Edison patented at the end of 1877 used tinfoil cylinders as the recording medium. Edison also sketched (but did not patent) recording devices using tape and disc recording media.

This phonograph cylinder dominated the recorded-sound market beginning in the 1880s. Lateral-cut disc records were invented by Emile Berliner in 1888 and were used exclusively in toys until 1894, when Berliner began marketing disc records under the Berliner Gramophone label. Berliner's records had poor sound quality; however, work by Eldridge R. Johnson improved the fidelity to a point where they were as good as cylinders. Johnson's and Berliner's companies merged to form the Victor Talking Machine Company, whose products would come to dominate the market for many years.

In an attempt to head off the disc advantage, Edison introduced the Amberol cylinder in 1909, with a maximum playing time of 4 minutes (at 160 RPM), which was in turn superseded by the Blue Amberol Record, whose playing surface was made of celluloid, an early plastic that was far less fragile than the earlier wax (in fact, it would have been more or less indestructible had it not been for the plaster of Paris core). By November 1918, the patents for the manufacture of lateral-cut disc records expired, opening the field for count-

70   7 Analog and Digital Sound Encoding

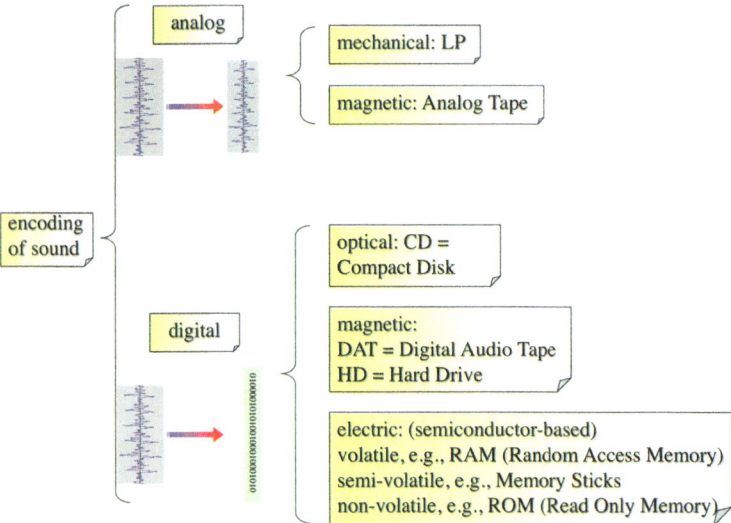

Fig. 7.5: Variants of analog and digital encoding.

less companies to produce them, causing disc records to overtake cylinders in popularity. Edison ceased production of cylinders in 1929. Disc records would dominate the market until they were supplanted by the compact disc, starting from the early 1980s.

7.3 The Digital Approach and Sampling

Philips and Sony have used digital encoding since 1982 on the recommendation of Herbert von Karajan. It is remarkable that his acceptance of CD quality (characterized by a 20 kHz upper frequency limit in the Fourier spectrum) was decisive for these companies, although Karajan's age (over 60 at the time) was not ideal as a reference for faithful sound perception. The hardware display of a CD is shown in Figure 7.7. The CD's laser reads the land and pit levels engraved upon the polycarbonate carrier as a bit sequence of zeros and ones.

The transformation from the analog soundwave to the digital representation on the CD is explained in Figure 7.8. The wave is quantized in two ways: First, the sound's amplitude is quantized by 16 Bits[1]. This means that we are given $2^{16} = 65{,}536$ values $(0, 1, 2, \ldots 65{,}535)$ defined by the binary integer representation $b_{15}2^{15} + b_{14}2^{14} + \ldots b_1 2^1 + b_0 2^0, b_i = 0, 1$. Second, the (quantized) values of the wave are only taken every 1/44,100th of a second, i.e., the sample rate is 44,100 samples per second. In total, this gives 635 $MByte$, 1 $Byte$ = 8 Bit, CD capacity for a one-hour stereo recording. We shall see

---
[1] A one-Bit quantization would allow for just two values, 0 and 1.

7.3 The Digital Approach and Sampling 71

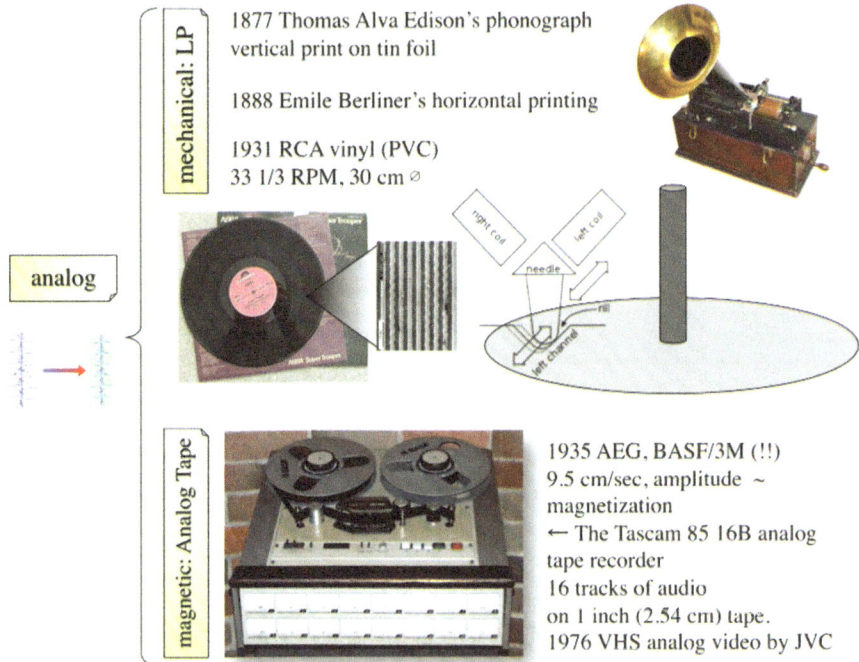

Fig. 7.6: Classical analog encoding and its historical highlights.

in the next subsection that this allows for partials up to frequency $44{,}100/2 = 22.05\ kHz$. The human ear is known to perceive up to 20 kHz.

Karajan, in his late sixties, could very probably not hear more than 15 kHz, so for him the quality of the CD was perfect. A young human however would have asked for higher resolution. In recent technology, a sample rate of 96 kHz with amplitude quantization of 24 Bit is being envisaged.

This looks like the endpoint of a long development of sound conservation and transfer, for which the 200 billion CDs sold by 2007 is a good argument. However, the creative argument comes from the critical concept: the container unit of music, the CD. First, its material part, the disc, why should this be the container? And second, why should music be transferred using such a hardware container? Because the Internet has defined the global reality of information transfer since the early 1990s, the answer to the above questions has become straightforward: Of course music data transfer can be accomplished via theInternet.

This transfer uses the electromagnetic digital technology that we display in Figure 7.9. And the unit of transfer is no longer a collection, such as those 635 $MByte$ (M stands for one million) containing a small number of musical compositions. This answer is, however, only theoretically valid because those 635 $MByte$ are far too much to be transferred in reasonable time with the

72   7 Analog and Digital Sound Encoding

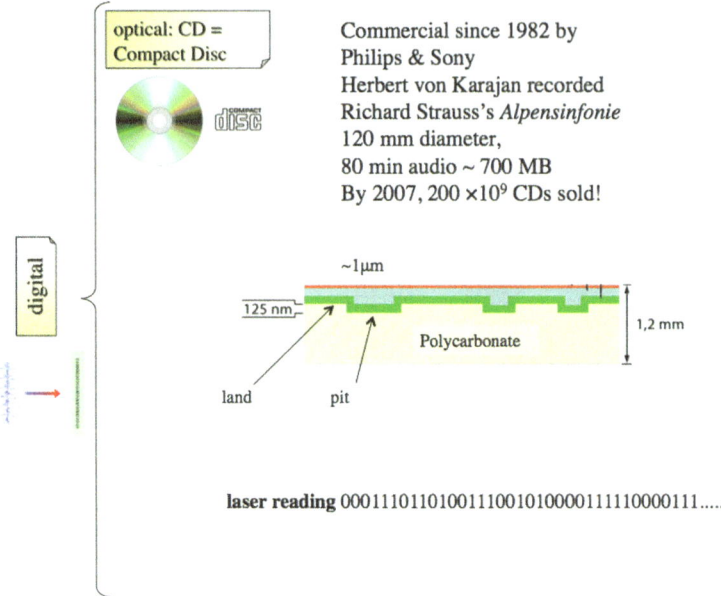

Fig. 7.7: The hardware display of a CD. The CD's laser reads the land and pit levels engrained upon the polycarbonate carrier as a bit sequence of zeros and ones.

present Internet performance. We shall see in the following sections that the solution of this performance problem has been an odyssey from the rejection of patents and the uninterest of industry to the celebrated integration of new ideas in the now commonly known MP3 format.

7.3 The Digital Approach and Sampling

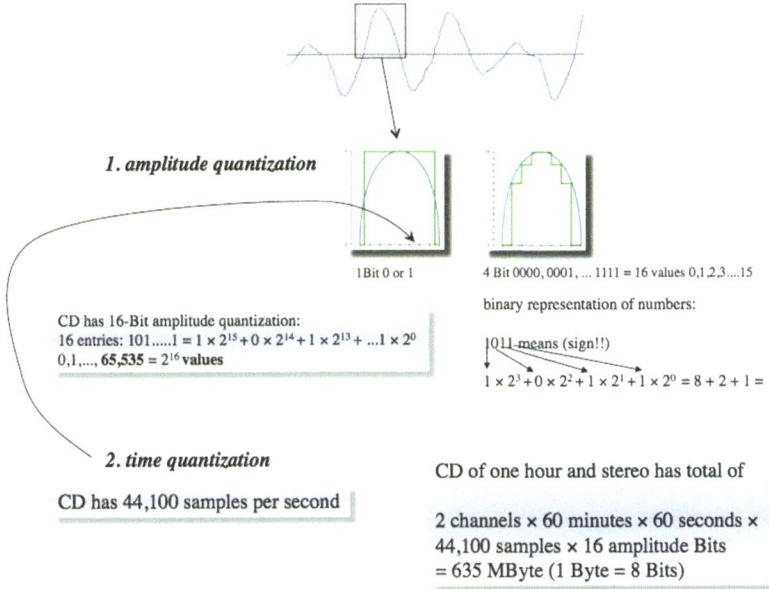

Fig. 7.8: The transformation from the analog soundwave to the digital representation on the CD.

7 Analog and Digital Sound Encoding

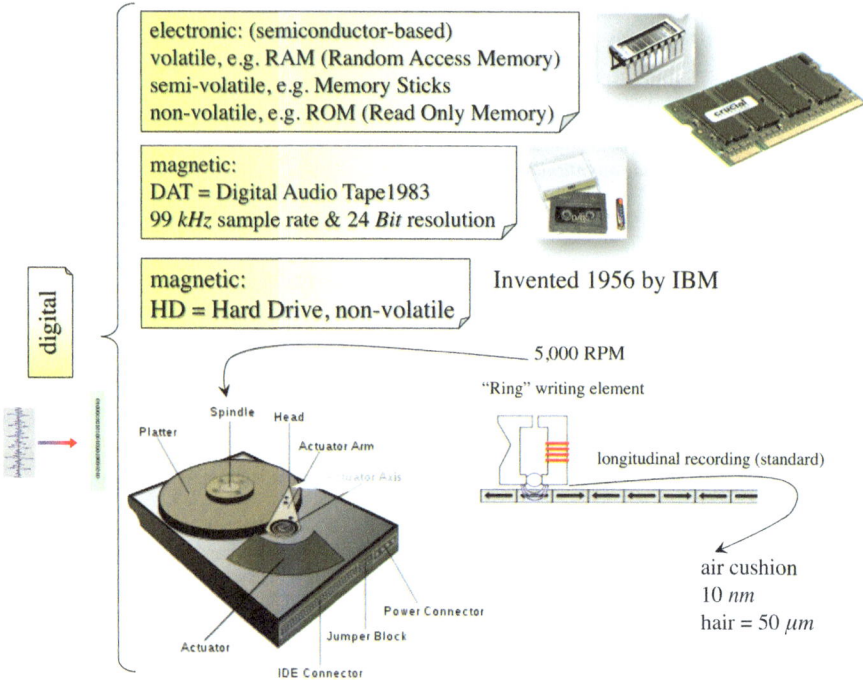

Fig. 7.9: Transfer modes of analog sound using electromagnetic digital technology.

8
Finite Fourier

Summary. Following the brief introduction to Fourier theory in Chapter 4.1, we will now discuss finite Fourier analysis, which confines the periodic Fourier formula to a finite number of overtones. We further introduce Nyquist's Sampling Theorem, which connects the highest overtone frequency with the sample rate. This chapter also introduces the Fast Fourier Transform (FFT), which increases the calculation efficiency of Fourier's formula by using complex numbers. Later in this chapter we talk about compression technologies, audio formats in various resolutions, and how they differ.

$$-\Sigma-$$

8.1 Finite Fourier Analysis

Let us first recall from Chapter 4.1.1 the Fourier formula for a sound function $w(t)$ of frequency f and look at its structure:

$$w(t) = A_0 + \sum_{n \geq 1} A_n \sin(2\pi \cdot nft + Ph_n).$$

It has two drawbacks:

1. In real life, a (non-zero) function is never strictly periodic. In particular, it cannot last forever, and we cannot know it except for a finite time interval.
2. It is not possible to recognize the function's value at every time; this would be an infinite task.

In real life we deal with functions that have a finite interval of definition, and we can measure values for only a finite number of times. This is, of course, not what must be offered to a Fourier setup. The solution of this situation is as follows. We are given a function $w(t)$ that is only defined in a finite interval of length P (see Figure 8.1). No periodicity is presupposed for this function—it might be a short sound recording of 0.5 seconds. To make these data amenable

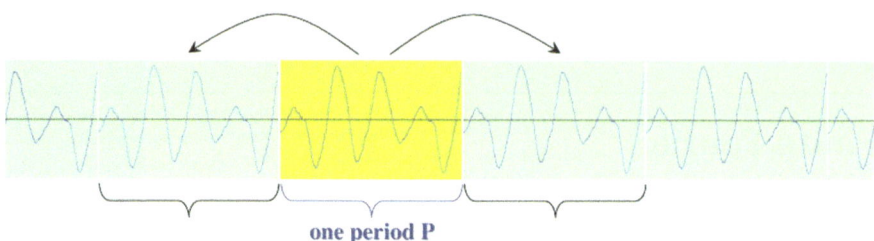

Fig. 8.1: To make general sound amenable to Fourier analysis, we have to create a periodic function. We continue the function by copying it infinitely to intervals of same length P to the left and right of the original interval such that the period of this continuation is P by definition.

to Fourier analysis, we have to create a periodic function of indefinite length from $w(t)$. We continue the function by copying it infinitely to intervals of the same length P to the left and right of the original interval such that the period of this continuation is P by definition. $w(t)$ is certainly not periodic, but we are not interested in anything else but the representation of this function in its original domain of definition. So whatever we have outside this does not matter. And the solution is to take the continuation of the function by shifted copies, so the result is periodic and coincides with the given function in its original domain.

The second problem is dealt with by measuring only finitely many values of $w(t)$ in the original interval (see Figure 8.2). It is related to a technically reasonable procedure where the function values are measured at regular time periods Δ. This Δ is called the *sample period*. The frequency $1/\Delta$ is called the *sample frequency*. The total sample period P is a multiple of the sample period, $P = N\Delta = 2n\Delta$, which means that we assume the multiplicity of the sampling action is even, $N = 2n$. This will be important later in Section 8.2, when we discuss Fast Fourier calculation.

So, we have fabricated a periodic function with frequency $f = 1/P$, and we have to find a Fourier representation that coincides with all the values measured in the $2n$ sample times. This is a mathematical problem relating to a Fourier formula that satisfies the preceding conditions. To solve this problem, we use the trigonometric equation

$$a \cdot \cos(x) + b \cdot \sin(x) = A \cdot \sin(x + \arccos(\frac{b}{A})), \text{where } A = \sqrt{a^2 + b^2}.$$

Using this equation in the Fourier equation for the overtone expression

$$A_m \cdot \sin(2\pi m f t + Ph_m),$$

we can write the Fourier equation at time $t = r\Delta$:

8.1 Finite Fourier Analysis

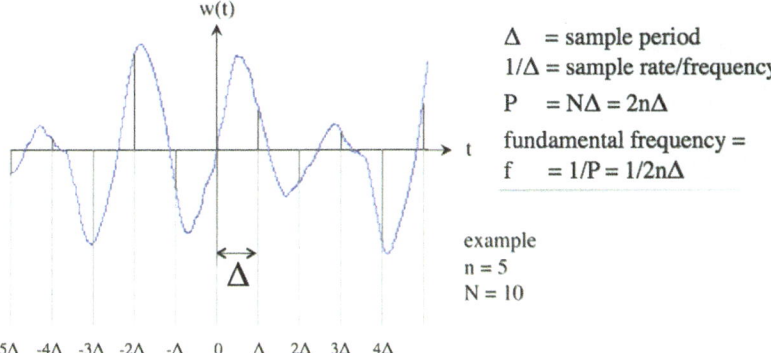

Fig. 8.2: Measuring only finitely many values of $w(t)$ in the original interval.

$$w(r\Delta) = a_0 + \sum_{m=1,2,3,\ldots n-1} a_m \cdot \cos(2\pi m f r\Delta) + b_m \sin(2\pi m f r\Delta) + b_n \sin(2\pi n f r\Delta).$$

The $2n$ equations for $r = -n, \ldots n-1$ are linear in the $2n$ unknown coefficients a_0, a_m, b_m, b_n. It can be shown, using the so-called orthogonality relations of sinusoidal functions [38, Section 38.8], that these equations have exactly one solution. This means that we can find exactly one Fourier formula that solves these conditions. The formula has its maximal overtone at frequency

$$nf = \frac{Nf}{2} = \frac{1}{2\Delta},$$

which is half the sampling frequency. This result is the content of the following theorem:

Theorem 1 (Nyquist's Sampling Theorem) *Given a wave $w(t)$ defined in a period P of time as above, if we measure its values every sample period Δ, dividing the period P into $N = 2n$ intervals, then there is exactly one set of coefficients a_0, a_m, b_m, b_n in the above finite Fourier expression for all sample times $r\Delta$, $r = -n, \ldots n-1$. The maximal overtone of this expression has index $n = N/2$, half the sample index N. Its frequency is half the sample frequency $nf = Nf/2 = 1/(2\Delta)$; it is called the* Nyquist frequency.

This result applies to CD sampling at $44.1 \ kHz$. The Nyquist frequency is half of this, $22.050 \ kHz$, as already announced above. Let us give another simple example: Suppose we have a wave of period $P = 20$ seconds and a sample frequency of $80 \ Hz$. This means that $\Delta = 1/80$ seconds and $N = 80 \times 20 = 1,600$. The fundamental frequency is $f = 1/P = 0.05 \ Hz$. How many overtones do we have in the finite Fourier analysis? We start with the fundamental at f, and go to the overtone with index $1,600/2 = 800$, having frequency $800 \times 0.05 = 40 \ Hz$.

The Nyquist theorem is the connection between general mathematical Fourier theory using infinite sums and infinite information about the waves, and the practical theory that is based upon finite information about the limited extension of the non-periodic wave function in time and the finite number of practically possible samples of this function. This finite Fourier theory solves the conflict described at the beginning of this section. But it does not solve the problem of the time needed to calculate the Fourier coefficients if the sample number N is large. For example, if we are given the CD sampling rate of 44,100 per second, and if we have a period of $P = 10$ minutes, we get $44,100 \times 600 = 2.6460 \times 10^6$ equations, and the calculation of their solution involves on the order of the square 7.001316×10^{12} of arithmetical operations. This is beyond reasonable calculation power. In the next section, we shall solve this major problem.

8.2 Fast Fourier Transform (FFT)

Fast Fourier Transform (FFT) is an algorithm that enables faster calculation of Fourier coefficients in the finite Fourier theory described above. This approach however is not easily accessible by the classical description of Fourier's formula using sinusoidal functions. We have to prepend a short discussion of the use of complex numbers to restate the formula. Then we shall be able to understand the beauty of FFT.

8.2.1 Fourier via Complex Numbers

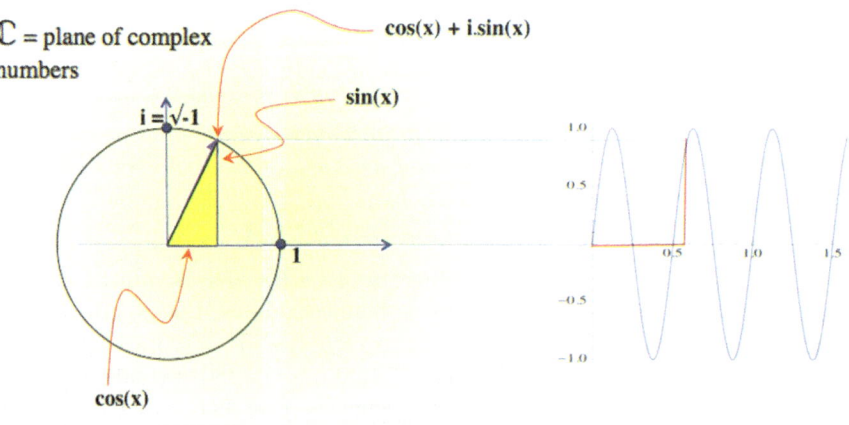

Fig. 8.3: The complex representation of sinusoidal functions.

8.2 Fast Fourier Transform (FFT)

In the discussion of the Nyquist theorem, we have restated the Fourier repesentation of the periodic wave function $w(t)$ in the shape

$$w(t) = a_0 + a_1 \cos(2\pi.ft) + b_1 \sin(2\pi.ft) + a_2 \cos(2\pi.2ft) + b_2 \sin(2\pi.2ft) + \ldots$$

This representation must be replaced by a much more elegant representation involving complex numbers. Recall that a complex number is just a vector $x = a+i.b$ in the plane $\mathbb{R}^2$ of real numbers (with coordinates $a, b \in \mathbb{R}$), where $i = \sqrt{-1}$, and besides the usual coordinate-wise addition, we have multiplication defined by $(a + i.b) \cdot (u + i.v) = (au - bv) + i(av + bu)$, using $i^2 = -1$. This enables a complex representation of sinusoidal functions as shown in Figure 8.3. The idea is to represent the sinusoidal values as the two coordinates of an arrow of length one rotating around the origin, where time is parameterized by the angle of the arrow with the x axis. The deeper advantage of this reduction of sinusoidal functions to rotating arrows is that this arrow's head satisfies Leonhard Euler's famous equation

$$\cos(x) + i.\sin(x) = e^{i.x}$$

where $e^z = \sum_{n \geq 0} \frac{z^n}{n!}$ is the exponential function. The advantage of this exponential representation is that the exponential function has a beautiful property: namely, it transforms addition of arguments into the product of values: $e^{z+w} = e^z \cdot e^w$. This is the basis of all apparently complicated formulas for sinusoidal functions. The creative idea behind this Eulerian result is (1) to connect $\cos(x)$ to $\sin(x)$ in the complex vector, and (2) to use complex arithmetic for the combination of such complex vectors.

The relationship between sinusoidal functions and the exponential is this:

$$\cos(x) = (e^{ix} + e^{-ix})/2$$
$$\sin(x) = (e^{ix} - e^{-ix})/2i.$$

This is used to restate the Fourier formula in terms of exponential expressions as follows:

$$w(t) = \sum_{n=0,\pm 1,\pm 2,\pm 3,\ldots} c_n e^{i2\pi.nft}$$

with the new coefficients c_n relating to the old ones by

$$a_0 = c_0,$$

and for $n > 0$,

$$a_n = c_n + c_{-n}$$
$$b_n = i(c_n - c_{-n}).$$

With this data, the above finite Fourier equation for $w(r\Delta)$ looks as follows: In order to keep ideas transparent, we only look at the special case of a sound sample of period $P = 1$ from $t = 0$ to $t = 1$, such that $\Delta = 1/N$ and $r\Delta = r/N$ for $r = 0, 1, \ldots N - 1$. We may then write

$$w_r = w(r\Delta) = w(r/N) = \sum_{m=0,1,2,3,\ldots N-1} c_m e^{i2\pi.mr/N}.$$

Why are the negative indices gone here? They are not, but they are hidden in the equation

$$e^{i2\pi.mr/N}.e^{i2\pi.m(N-r)/N} = e^{i2\pi.mr/N+i2\pi.m(N-r)/N} = e^0 = 1,$$

which means

$$e^{i2\pi.m(N-r)/N} = e^{i2\pi.(-m)r/N},$$

where $-m$ is a negative index! Also, c_{N-m} is the complex conjugate of c_m since a_m, b_m are all real numbers. Therefore we have a total of $N/2$ independent complex coefficients, i.e., N real coefficients as required from the original formula.

This enables us to represent the sample sequence with the vector

$$w = (w_0, w_1, \ldots w_{N-1}) \in \mathbb{C}^N$$

in the N-dimensional complex vector space $\mathbb{C}^N$. In this space, we have a canonical scalar product of such vectors u, v (similar to the high school formula)

$$\langle u, v \rangle = \frac{1}{N} \sum_{r=0}^{N-1} u_r \overline{v_r}.$$

We have exponential functions

$$e_m = (e_m(r) = e^{i2\pi.mr/N})_{r=0,1,2,\ldots N-1}$$

and their scalar product is

$$\langle e_m, e_m \rangle = 1,$$
$$\langle e_m, e_q \rangle = 0, m \neq q,$$

which means that they define an orthonormal basis of $\mathbb{C}^N$ (see Figure 8.4). These relationships are the orthogonality relations mentioned above. These functions replace the sinusoidal functions in the complex representation.

Every sound sample vector $w = (w_0, w_1, \ldots w_{N-1})$ can be written as a unique linear combination (see Figure 8.4)

$$w = \sum_{m=0,1,2,3,\ldots N-1} c_m e_m,$$

and the coefficients c_m are determined by the formula

$$c_m = \langle w, e_m \rangle = \frac{1}{N} \sum_{r=0,1,2,3,\ldots N-1} w_r e^{-i2\pi.mr/N}.$$

8.2 Fast Fourier Transform (FFT)   81

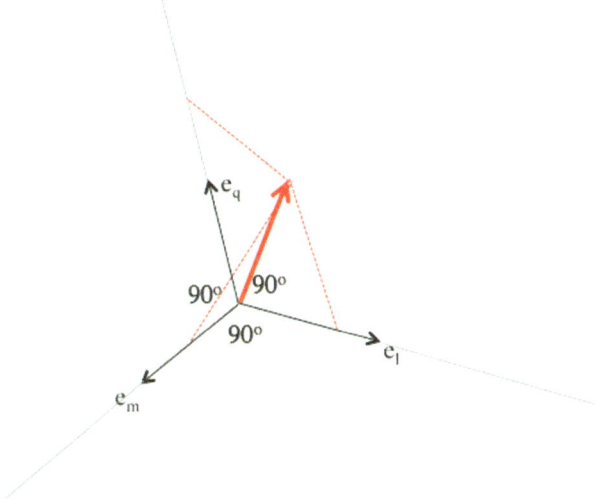

Fig. 8.4: An orthonormal basis of $\mathbb{C}^N$.

This writing in the complex vector space, using the exponential basis, is the *Fourier theorem for such finite sums*. It is much more elegant than the original one using sinusoidal waves, and the calculation of the Fourier coefficients (essentially the amplitude and phase spectra) is very easy. We hope that the reader can appreciate this beautiful simplification of the theory through the step to the complex numbers.

8.2.2 The FFT Algorithm

The FFT algorithm was invented to calculate the Fourier coefficients c_m more quickly than in the quadratic growth algorithm that is given by the solution of the above orthogonality relations. The FFT was discovered by James W. Cooley (IBM T.J. Watson Research Center) and John W. Tuckey (Princeton University and AT&T Bell labs) and published in 1965 in a five-page (!) paper, *An Algorithm for the Machine Calculation of Complex Series* [12]. It has become one of the most-cited papers of all time. It is remarkable that the algorithm was already discovered by the great German mathematician Carl Friedrich Gauss in 1805 for astronomical calculation, but it was not recognized because he wrote the paper in Latin. (This might be amusing, but it also teaches us to share creations in languages that are communicative. It is interesting that many great creators do not care about this and even avoid any straightforward communication of their findings. The psychology of such hidden creativity has not been written about to date.) Compared to the quadratic growth N^2 defined by the orthogonality relations, this FFT algorithm grows with $N.\log(N)$.

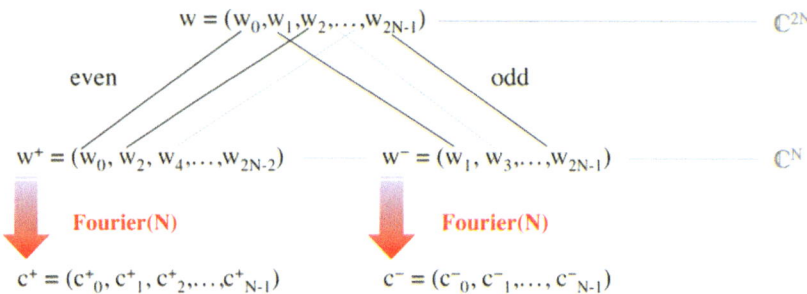

Fig. 8.5: The FFT trick: split the vector $w = (w_0, w_1, \ldots w_{2N-1}) \in \mathbb{C}^{2N}$ into its even and odd parts w^+, w^-.

The idea is very simple but ingenious, see Figure 8.5. We now suppose that N is not only even, but a power of 2. The trick consists of taking the sample vector $w = (w_0, w_1, \ldots w_{2N-1}) \in \mathbb{C}^{2N}$ and splitting it into its even and odd parts w^+, w^- as shown in Figure 8.5. Supposing that we have calculated the coefficient vectors $c^+ = (c_0^+, c_1^+, c_2^+, \ldots c_{N-1}^+), c^- = (c_0^-, c_1^-, c_2^-, \ldots c_{N-1}^-)$ with a growth $Fourier(N)$, we can calculate the coefficient vector c of w using the formula $c_r = (c_r^+ + e(N)^r . c_r^-)/2$. This latter formula permits the calculation $Fourier(2N) \leq 3 \times 2N + 2N + 2.Fourier(N) = 2.Fourier(N) + 8N$. This is a recursive formula shows that the growth can be set to that of $N. \log(N)$.

The detailed mathematics here is not so important; the question is rather how this problem can be solved in such a simple way. Nothing suggests that this formula consists of sub-information that can be calculated separately and yield the desired values. Once we are told that such a procedure is possible, the solution is not difficult. The solution is just the splitting of the wave w into its even and odd parts. Then, once the even and odd coefficients have been calculated, the connecting formula $c_r = (c_r^+ + e(N)^r . c_r^-)/2$ immediately yields the required result. So the creative part of this work is not the mathematics, but the unbelievable fact that everything can be solved recursively from even and odd parts.

8.3 Compression

In order to appreciate the quality of MP3 compression, we have to briefly discuss the concept of compression in computer file formats for digitized audio data.

Such formats contain not only the audio-related data, but additional technical information about the encoding method of the raw data, and data concerning the composer and other poietic information. This additional information can be complex depending on the actual method. We shall see for MP3 that the encoded data stream is in itself a virtuosic achievement invented to enable

8.3 Compression

a piece to be played starting at any time moment during the piece.[1] The additional data strongly depend on compression methods invented to reduce the saved data with respect to the original audio data.

By definition, compression for a determined file format is the ratio of two quantities: original data/file data.

For example, if the original data is a 10-minute CD and the file needs to be 10 MB, what is the compression? Answer: 44.1×10^3 $Hz \times 16$ $Bit \times 2$ channels $\times 10$ $min \times 60$ $sec/10^7 \times 8$ $Bit \approx 10.58$.

There are two types of compression:

- *Lossless compression.* This means that the original data can be completely reconstructed from the compressed data.
- *Lossy compression.* This means that the original data cannot be completely reconstructed from the compressed data.

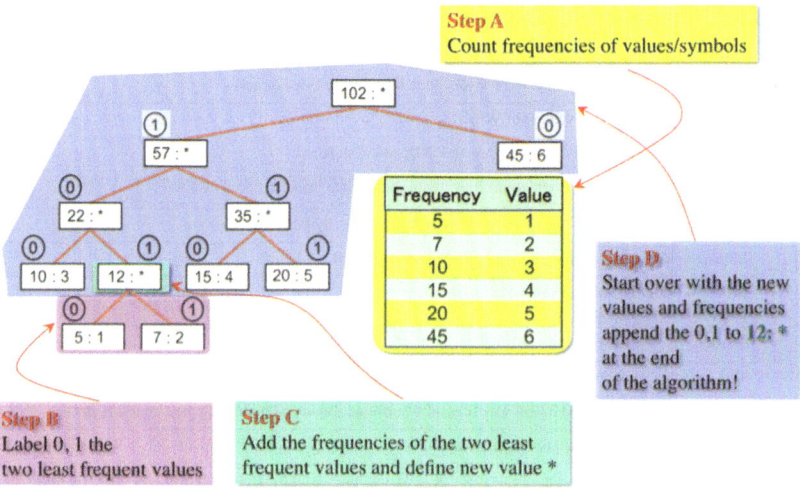

Fig. 8.6: An example of Huffman compression.

Here are two lossless compression methods also used in MP3:

- *RLE = Run Length Encoding.* IThis consists of replacing a Bit sequence (such as black and white pixels in a black-and-white TV screen) with the sequence of numbers of uninterrupted occurrences of a determined Bit.
- *Huffman.* This method is described in Figure 8.6. We start with a number of values, here six (step A), and list their frequencies. Label the two least frequent values 0,1 (step B), then add the frequencies of the two least frequent values and define a new value * (step C). Start over with the new

---

[1] Remember that Fourier's coefficient calculation needs all time moments!

values and frequencies, and append the 0 and 1 to 12:* at the end of the algorithm (step D).

And here are two examples of lossy compression:

- *Analog to Digital Conversion (ADC).* From analog to CD with 44,100 Hz and 16-*Bit* amplitude digitization. This has an infinite rate.
- *Quantization.* In the 16-*Bit* representation of amplitude, the last five Bits are neglected. So, for example, 1011 1100 0101 1110 is transformed in to 1011 1100 0100 0000 and then shortened to the 11 relevant Bits remaining, 1011 1100 010.

8.4 MP3, MP4, AIFF

Figure 8.7 shows the best-known digital audio file formats, together with their compression types.

uncompressed		WAV	Microsoft (extends RIFF = Resource Interchange File Format)
		AIFF	Mac (Mac's version of WAV)
		AU	Simple SUN audio file format, UNIX sound standard
		PCM	(Pulse Code Modulation) Raw digitao audio data, usually stored in WAV
compressed	lossless	ALAC	= Apple Lossless, stored in MP4 container, e.g., in iPods
		Monkey's Audio, better than Shorten or FLAC and WavPack	
		MPEG-4 (multimedia) SLS (Scalable Lossless), extended to AST (Audio Lossless), DST (Direct Stream Digital) based on Apple's QuickTime	
		Shorten, used to compress CD files	
		TTA (True Audio) compression on multichannel 8, 16, and 24 Bit data of wav files, GPL, cross-platform	
		WavPack 8, 16, 24, and 32 Bit wav files, also surround sound and high frequency sampling rates	
		WMA Lossless developed for archival purposes	
	lossy	AAC (advanced audio coding) a successor of MP3, complies with MPEG-2 and MPEG-4, up to 96 kHz	
		ATRAC (adaptive transform acoustic coding) Sony, developed for MiniDisc 1992	
		MP3, to be discussed	
		Musepack, also known as MP+, for PC, Linux, and Mac OSX, violates patents from MP3, similar to MP3	
		Ogg Vorbis, popular in free software circles	
		WMA (Windows Media Audio), part of the Windows Media framework, originally competitor to MP3	

Fig. 8.7: The best-known digital audio file formats, together with their compression types.

Let us discuss MP3, the most successful format. MPEG (Moving Pictures Experts Group) is the code name for the standardization group ISO/IEC JTC1/SC29/WG11 (Int. Organization for Standardization/International Electrotechnical Commission). It was created in 1988 to generate generic standards

8.4 MP3, MP4, AIFF

for encoding digital video and audio data. MPEG-1 is the result of the first work phase of the group and was established in 1992 as ISO/IEC IS 11172 standard. It contains Layer-1, Layer-2, and Layer-3, which means three operation modes with increasing complexity. By MP3 one denotes Layer-3 of MPEG-1. MPEG-2 Advanced Audio Coding (AAC) is the result of the second work phase. It enhances Layer-3 in many details. We are not going to discuss this phase here.

The development of these compressed audio formats goes back to research by Dieter Seitzer since 1960, who at that time worked at IBM, and his student Karlheinz Brandenburg, who is above all responsible for the psychoacoustical compression methods. It is remarkable that Seitzer's patent was rejected in 1977; however it was awarded in 1983 but was then suspended because of lack of interest from the industry! MP3 is above all based on research and development by Brandenburg at the Fraunhofer Institut für Integrierte Schaltungen (IIS) in Erlangen, Germany. It is an open standard, but it is protected by many patents (more than 13 US patents, and more than 16 German patents). We shall discuss legal aspects at the end of this topic.

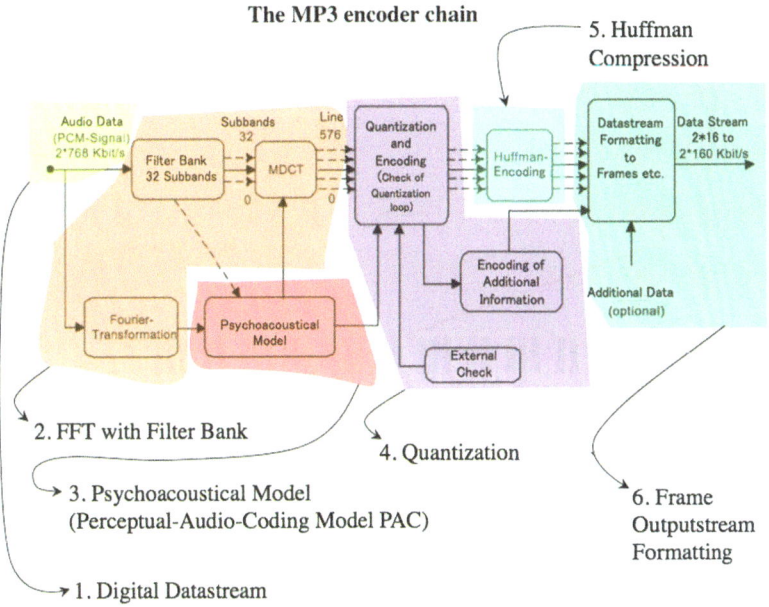

Fig. 8.8: The MP3 encoder chain with its core Perceptual-Audio-Coding device.

MP3 includes these options:

- Mono and stereo, in particular joint stereo encoding for efficient combined encoding of both stereo channels.

- Sampling frequencies include 32 kHz, 44.1 kHz, 48 kHz; for MPEG-2 also 16 kHz, 22.5 kHz, 24 kHz; and for MPEG-2.5 (Fraunhofer-internal extension) also 8 kHz, 11.5 kHz, and 12 kHz.
- The compression's bitrate (= the Bits traversing the audio file per second) goes from 32 $kBits/sec$ (MPEG-1) or 8 $kBits/sec$ (MPEG-2) up to 320 $kBits/sec$. For MP3 the bitrate can even vary from frame to frame (a frame is the unit package in MP3; we will come back to it later) and, together with the so-called bit-reservoir technology, allows variable as well as constant bitrates.

Besides compression, MP3 has the advantage of being platform-independent. This is also a strong reason for its popularity.

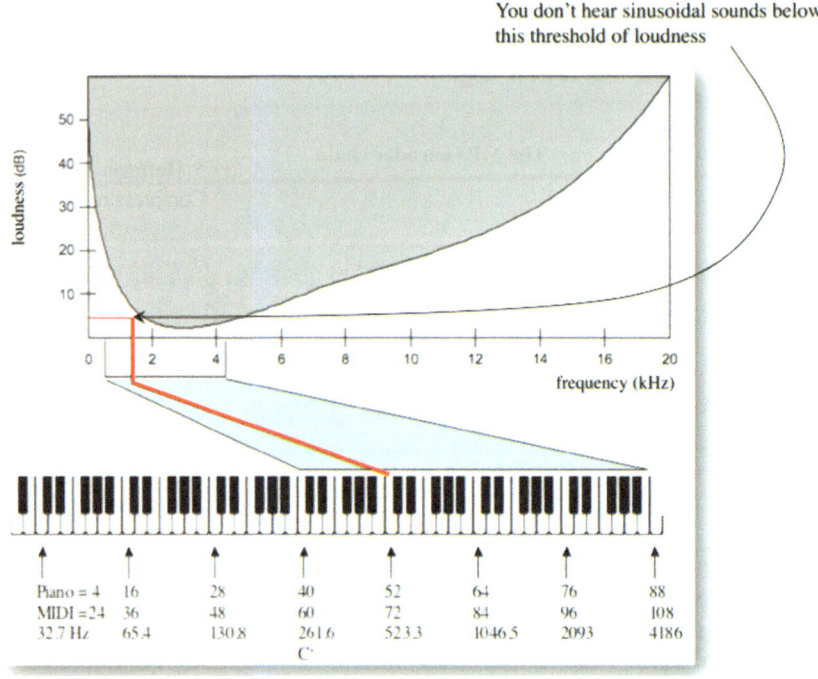

Fig. 8.9: Hearing thresholds: One uses the fact that there is a loudness threshold below which the human ear does not hear sounds.

Let us now look at the MP3 encoder chain (the decoder works symmetrically; we don't discuss it for this reason). Figure 8.8 shows the original overall flowchart of audio information. 1. From the left, the digital data stream defines the system's input; the quantity is given by 768 $kBit/sec \approx 48{,}000 \times 16\ Bit/sec$. In region 2, FFT processing of the digital signal is applied, yielding a number of Fourier coefficient banks (filter banks—we discuss below what is a filter, but

8.4 MP3, MP4, AIFF

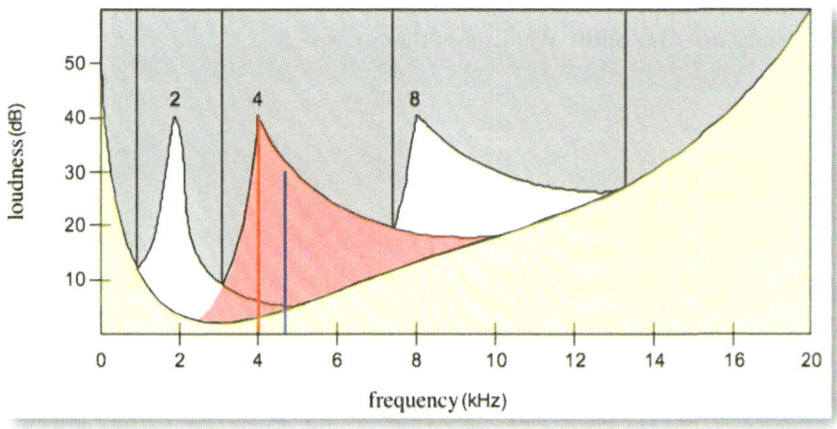

Fig. 8.10: Auditory masking uses the fact that for a given frequency component, there is a curve below which no other component can be heard.

essentially it just selects a range of Fourier coefficients). We cut the spectrum $0 - 20\ kHz$ into 32 subbands of $625\ Hz$ each ($32 \times 625 = 20{,}000$) for $1/40\ sec$ windows. We then use MDCT (Modified Discrete Cosine Transformation, a variant of FFT) to split each $625\ Hz$ band into 18 subbands with variable widths, according to psychoacoustical criteria. We then get $576 = 18 \times 32$ "lines." Region 3 is the heart of the MP3 compression, covering 60 percent of the compression. It operates by elimination of physiologically superfluous information following the so-called Perceptual-Audio-Coding Model (PAC). PAC has three components: PAC 1—hearing thresholds; PAC 2—auditory masking; PAC 3—temporal masking.

PAC 1, hearing thresholds, uses the fact that there is a loudness threshold below which the human ear does not hear sounds. The curve is shown in Figure 8.9. Therefore we may eliminate Fourier components (overtones) with amplitudes below these thresholds.

PAC 2, auditory masking, uses the fact that for a frequency component there is a curve below which no other component can be heard. Figure 8.10 shows the situation for 2, 4, and 8 kHz components.

PAC 3, temporal masking, uses the fact that for every sinusoidal frequency component of frequency f and loudness l, another subsequent component cannot be heard below the given curve of loudness in time, because the ear needs some time to 'recover' from that first component's perception. This is even true before (!) the given one, because the perception needs to be built up. Figure 8.11 shows the curve.

In points 4 and 5 of the flowchart, we have the lossy quantization and lossless Huffman compressions discussed above—they make up the remaining

40 percent of compression. The final output is shown in region 6, where the MP3 frames are also built. We discuss them now.

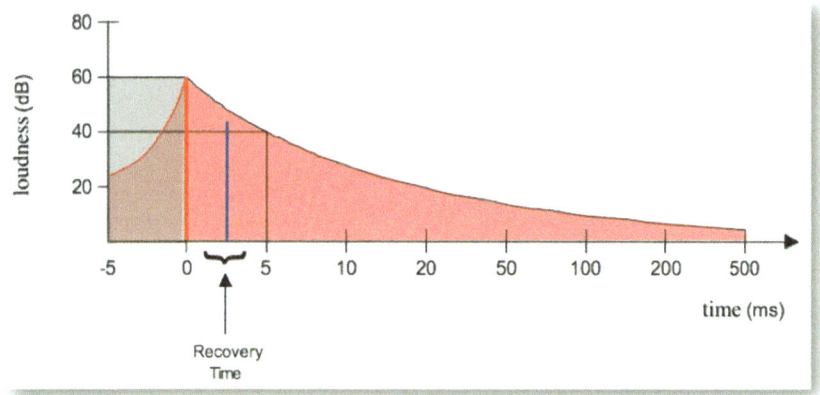

Fig. 8.11: Temporal masking uses the fact that for every sinusoidal frequency component of frequency f and loudness l, another subsequent component cannot be heard below the given curve of loudness in time, because the ear needs some time to 'recover' from that first component's perception.

To begin with, the MP3 file is headed by an MP3 file identifier, the ID3 tag. It requires 128 $Byte$ and its information is shown in Figure 8.12.

Bytes	Content
3	Tag = identification as ID3 tag
30	title of piece
30	name of interpreter(s)
30	name of album
4	year of publication
30	comment
1	genre identification

Fig. 8.12: The MP3 file identifier, the ID3 tag, requires 128 $Byte$.

After this header, a number of frames are added. Here is the concept: A frame is an autonomous information package. This means that all encoding data is provided within every frame to enable a file to be played from any given time onset. A frame's duration is $1/38.28125 \approx 1/40 \ sec$. This enables virtually continuous playing for humans.
Each frame has these parts:
- a 32-Bit header indicating the layer number (1-3), the bitrate, and the sample frequency;

8.4 MP3, MP4, AIFF

- the Cycle Redundancy Check (CRC) with 16 Bits for error detection (without correction option) but frame repetition until the correct frame appears;
- 12 Bits for additional information for Huffman trees and quantization info;
- a main data sample block of 3,344 Bit for Huffman-encoded data.

32-Bit Frame Header

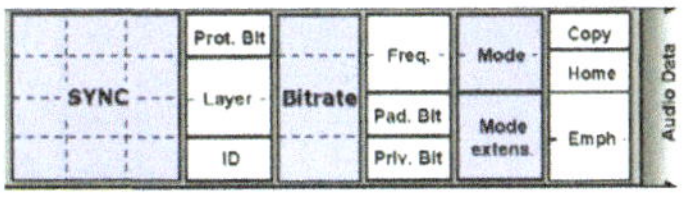

Position	Task	Length in bits
A	Frame-SYNC (for playing and "jumping around")	11
B	MPEG Audio Version (MPEG-1, -2, etc.)	2
C	MPEG Layer (Layer I, II, III, etc.)	2
D	Protection	1
E	Bitrate Index	4
F	Sampling Frequency (e.g., 44.1 kHz)	2
G	Padding bit (compensates incomplete allocation)	1
H	Private bit (application-specific trigger)	1
I	Channel mode (Stereo, Joint Stereo)	2
J	Mode Extension (for Joint Stereo)	2
K	Copyright	1
L	Original ("0" if copy, "1" if original)	1
M	Emphasis (outdated)	2

Fig. 8.13: Each frame has a 32-Bit header containing the information shown here.

Each frame has a 32-Bit header containing the information shown in Figure 8.13.

The frame structure is given as shown in Figure 8.14. Each frame has a header and a reservoir of $3,344$ Bit. However, the reservoir technique allows information to be placed in reservoirs that are not filled up yet. So data for block n might be saved in block $n+1$. This is called the *reservoir technique*.

There are two important formulas regarding frame capacities for MP3. The fixed data is this:

1. number of $frames/sec = 38.28125$
2. maximal audio data capacity per frame $= 3,344$ $Bit/frame$
3. number of frequency bands $= 32$

The first formula is this: The maximal bitrate

$$3,344\ Bit/frame \times 38.28125\ frame/sec = 128\ kBit/sec$$

guarantees CD quality.

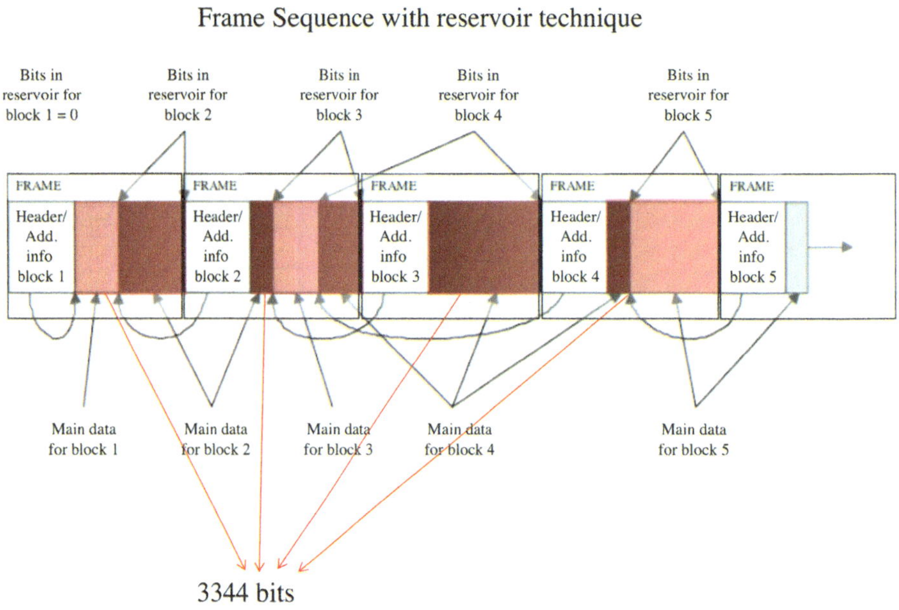

Frame Sequence with reservoir technique

Fig. 8.14: The MP3 frame structure.

And the second one is this:

$$(44{,}100\ time\text{-}sample/sec)/(38.28125\ frame/sec) =$$
$$1{,}152\ frequency\text{-}samples/frame$$

guarantees CD quality. This yields $1{,}152/32 = 36\ frequency\text{-}samples/band$. Observe: $625\ Hz/\text{band}/38.28125\ Hz = 16.3265\ frequency\text{-}samples/band$. We have overlapping data, but this is OK to minimize measurement errors.

Some performance values are shown in Figure 8.15.

Let us add some remarks regarding the joint stereo encoding: MP3 implements the Joint Stereo Coding compression method, which is based on these two principles:

- Mid/Side Stereo Coding (MSSC), where instead of taking the left and right channels (L, R), we use the equivalent data $(L + R, L - R)$—since L and R are usually strongly correlated, the difference is quite 'tame'.
- Intensity Stereo Coding (ISC), where the sum $L + R$ and the direction of the signal are encoded (replacing the $L - R$ information).

This coding method also uses the fact that the human ear is poor at localizing deep frequencies. Since the direction is detected by phase differences that are difficult to retrieve for deep frequencies, they are encoded mono! Let us end the discussion with some commercial/legal information: The license

8.4 MP3, MP4, AIFF

MPEG procedure	compression	quality	bitrate kBit/sec	bandwidth kHz	mode
MPEG-1 layer-3	14:1 – 12:1	CD	128	>15	stereo
MPEG-1 layer-3	16:1	Approximately CD	96-112	15	stereo
MPEG-2 layer-3	16:1-24:1	Radio quality	56-64	11	stereo
MPEG-2 layer-3	24:1	Language	32	7.5	mono
MPEG-2 layer-3	48:1	Shortwave radio	16	4.5	mono
MPEG-2.5 layer-3	96:1	Telephone	8	2.5	mono

Input bitrate (2 × 768) / output bitrate (128) = 12

Fig. 8.15: Some MP3 performance values.

rights of Fraunhofer IIS are represented by the French company Technicolor SA, formerly Thomson Multimedia. Here are the figures:

- USD 0.50 per decoder
- USD 5.- per encoder
- USD 15,000.- annual lump sum

This means that an enterprise that sells 25,000 copies annually of the encoder software, pays 25,000 × USD 5.- + USD 15,000.- = USD 140,000.- for the first year and then USD 15,000.- in annual fees for every successive year.

9
Audio Effects

Summary. This chapter reviews some important audio effects, such as filters, equalizers, reverberation, and time/pitch stretching.

$$- \Sigma -$$

9.1 Filters

In our discussion of finite Fourier methods, we had encountered the clever trick that transforms a non-periodic function in a time interval of length P (the period) into a periodic function by simply juxtaposing the shifted function to the left and right of the original function. This might be clever if we have a finite time interval of this function. But what happens to the function if it is extended to infinity and is not periodic? Then that method fails.

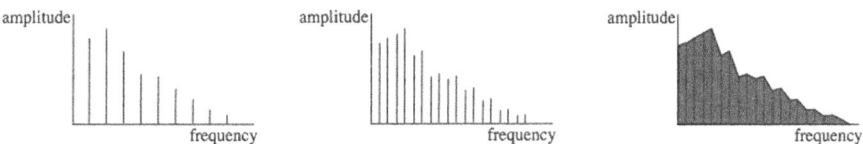

Fig. 9.1: Letting a finite frequency go to zero without losing the formal power of the original theory.

Mathematicians have however been able to solve this problem with a very adventurous idea: Why do we have to suppose a finite period for our theory? Is Fourier only feasible for finite periods? In other words, why couldn't we try to extend the theory to infinite periods? This means that we are looking for a Fourier formula for frequency $f \to 0$.

In our context we are fighting against the problem of letting a finite frequency go to zero without losing the formal power of the original theory (see Figure 9.1).

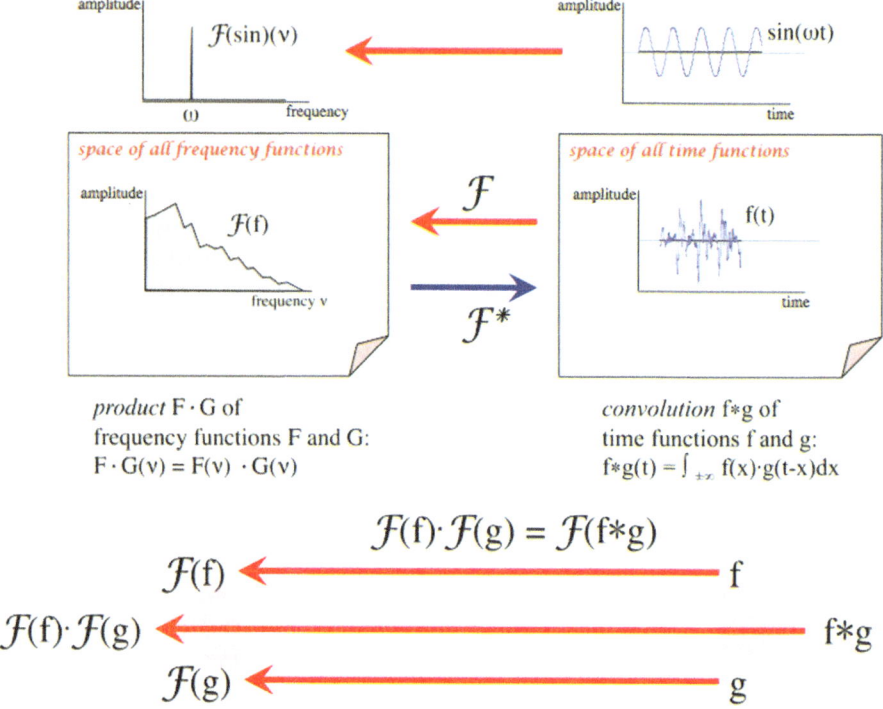

Fig. 9.2: The correspondence between the product of frequency functions and the convolution of their corresponding time functions.

The solution of this limit problem, the transition from a finite frequency to the vanishing frequency, works because the limit of a sum is, if mathematically correctly worked out, the integral, and the finite frequency converges to an infinitesimal frequency. Evidently, under this transformation, there will no longer be a discrete multiplicity of a fundamental frequency, but an entire continuous variety of frequencies that describe the Fourier formula. And here is the result. We use also the usual symbol $f(t)$ for the function (instead of $w(t)$), and ν for the frequency (instead of f):

$$\mathcal{F}(f)(\nu) = \frac{1}{\sqrt{2\pi}} \int_{-\infty}^{\infty} f(t) e^{-i\nu t} dt.$$

This function $\mathcal{F}(\nu)$ of frequency ν is called the *Fourier transform* of f. It is a function of frequency instead of time. Figure 9.2 gives an overview of the power of this new Fourier transform formula.

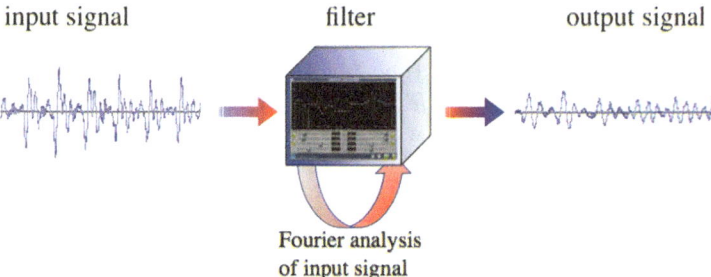

Fig. 9.3: A filter is a function that alters an audio signal on the basis of its Fourier transform.

Here are the properties: $\mathcal{F}$ defines a linear isomorphism from the space of all time functions (right) (with some technical properties that are not interesting in our context) onto the space of all frequency functions (left). The inverse of $\mathcal{F}$ is denoted by $\mathcal{F}^*$. For example, the sinusoidal function $\sin(\omega t)$ is transformed into the function that is 0 except at frequency $\nu = \omega$, where it is 1. So the sum of sinusodial functions is transformed into a sum of such spike functions, which symbolize the Fourier overtones with their amplitudes. Furthermore, by the inverse $\mathcal{F}^*$, the pointwise product $F \cdot G$ of two frequency functions F, G $((F \cdot G)(\nu) = F(\nu) \cdot G(\nu))$ is transformed into the *convolution* $(f * g)(t) = \int_{\pm\infty} f(x)g(t-x)dx$.

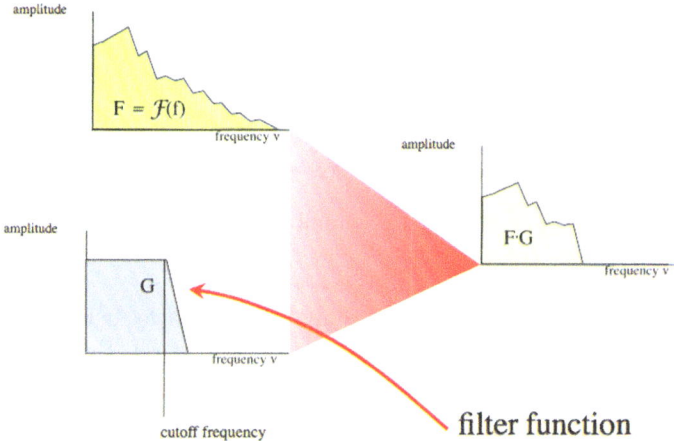

Fig. 9.4: The filtering process in the frequency domain for a time function $f(t)$.

This means that we can use pointwise products in the frequency domain to cut functions. This is precisely the fact we need to understand filters! *A filter*

9 Audio Effects

is a function that alters an audio signal on the basis of its Fourier transform (see Figure 9.3).

The filtering process for a time function $f(t)$ takes as input a Fourier transform $F = \mathcal{F}(f)$ of f and a *filter function* $G(\nu)$ in the frequency domain. The process is shaped in such a way as to eliminate a number of frequencies from $\mathcal{F}(f)$ by pointwise multiplication, yielding $F \cdot G$, and then to transform the result $F \cdot G$ back into the time domain, yielding $f_G = \mathcal{F}^*(F \cdot G) = f * \mathcal{F}^*(G)$ (see also Figure 9.4).

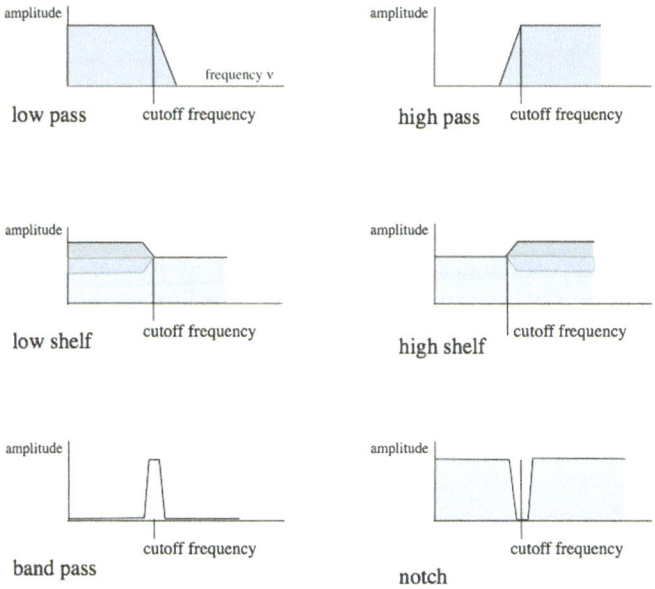

Fig. 9.5: Frequently used filter types.

In audio technology there are a number of special filters that are given standard names: low pass, high pass, low shelf, high shelf, band pass, notch. Their filter functions are shown in Figure 9.5.

The meaning of "low pass" is that only low frequencies below a determined cutoff frequency can "pass"; higher frequencies are cut off. By contrast, a low shelf filter passes all frequencies above a cutoff frequency, but increases or reduces frequencies below the cutoff frequency by a determined quantity.

9.2 Equalizers

In audio technology, the representation and management of filters is realized in two ways: as a graphic EQ and as a parametric EQ, which stands for "equaliza-

9.2 Equalizers

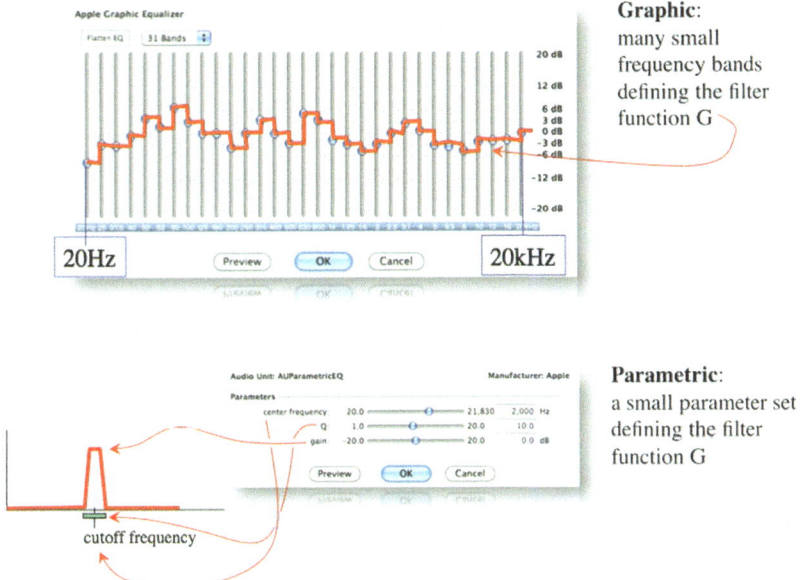

Fig. 9.6: Two types of equalizers, graphic and parametric, in commercial software.

tion" and just means filtering using complex filter functions. Figure 9.6 shows these two types in commercial software.

Let us conclude the topic of filters with a practical remark. Filters are not only used as abstract devices, but also occur in very practical contexts. Two simple examples are walls and electrical circuits (see Figure 9.7). If a sound is produced in a room, it propagates to the room's walls. There, it is reflected but also travels through the walls. The sound that arrives in the rooms adjacent to the room where the sound is produced is (hopefully) very different from the original one. The walls usually act as low pass filters. This means that the sound in the adjacent room(s) contains a certain amount of only low frequencies. This is made clear when we think of the bass sound in commercial music. In the adjacent rooms, you will (have to) hear the bass, an effect that can be very disturbing at times.

The second example, a special electrical circuit shown in the lower part of Figure 9.7, is capable of realizing a low pass filter in an electrical setup. It takes the input as a voltage V_{in} and 'filters' it using a resistance R and a capacity C. The capacity induces a low pass filter in that it only transmits high frequencies, and therefore the low frequencies pass at V_{out}.

The cutoff frequency is $1/2\pi RC$. It should be remarked that the use of electrical circuits to simulate acoustical configurations is a powerful standard method in acoustics, especially acoustics of musical instruments. It serves as a methodology of representation in acoustics, but also as a means for the construction of acoustical effects on the level of electrical sound hardware technology.

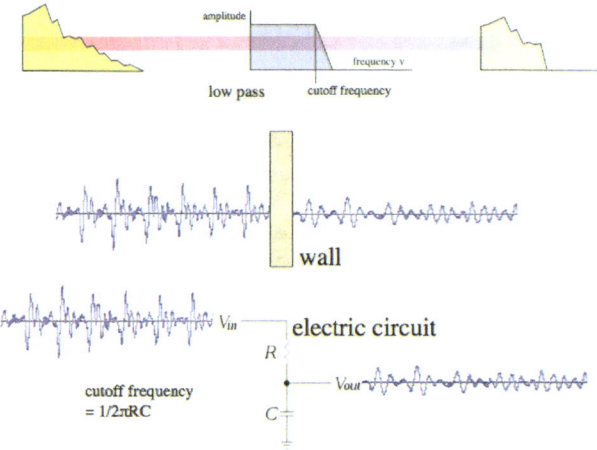

Fig. 9.7: A special electrical circuit shown in the lower part of the figure is capable of realizing a low pass filter with an electrical setup.

9.3 Reverberation

In physics and psychoacoustics, reverberation (commonly called reverb) describes how sound is reflected by the spatial environment. When a sound is produced, a listener hears the sound first as it comes directly from the source. The sound simultaneously travels in all directions, reflecting off objects or walls, possibly returning in the direction of the listener. Before the original sound ends, many of these reflections reach the listener (see Figure 9.8). Unlike an echo, no break in the sound is perceived. The shape and size of a given room affect these reflections.

People sometimes think of reverb as the aftertaste of a sound. These reflected waves add sustain and volume to the original sound. Reverb creates a sense of "spaciousness" and is sometimes perceived as a "divine" sound. However, it also diminishes the intelligibility of speech sound and makes it more difficult for people to locate the sound's source, sound example: reverb.mp3. In general, reverberation is most present in recital and orchestral halls. Acoustical engineers consider different musical and speech purposes when they design a room. Music that is performed in recital halls is not typically amplified and therefore relies on reverb to complete the sound (see Figure 9.9). When engineers design a live-music club, they don't worry about reverberation because the music is always amplified (see Figure 9.10).

9.3 Reverberation 99

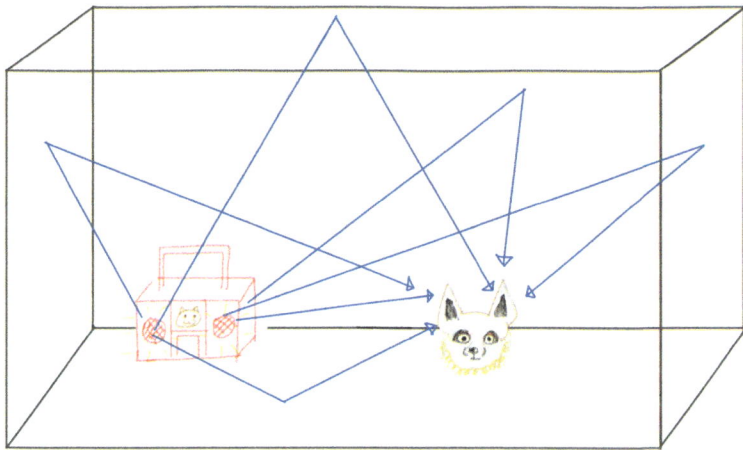

Fig. 9.8: Visualized reverberation in a typical room. A sound source (represented by the speaker) can use various routes to reach the ears (represented by the fox's ears).

Fig. 9.9: Orchestral Hall, Minnesota Orchestra, in Minneapolis.

Furthermore, in computer science, digital reverberators use signal-processing algorithms to simulate reverberation digitally (see Figure 9.11). Digital reverberation algorithms use several feedback delay circuits to add decay to the original sound, similarly to how real reverberation is formed by adding reflections, see [50, Chapter 2] for details. Moreover, digital reverb generators can not only simulate general reverberation effects, but also the time and frequency response of specific environments (bedroom, shower room, outdoor, orchestra

100   9 Audio Effects

Fig. 9.10: The Whole Music Club at the University of Minnesota, Minneapolis.

hall, etc.). This is done by simulating the environment dimensions and absorption factors to calculate the exact reverberation responses that would happen in that environment. Nowadays, reverb is one of the most commonly used effects in audio editing and is applied extensively in keyboard synthesizers, electric instrument pedals, and digital audio workstations.

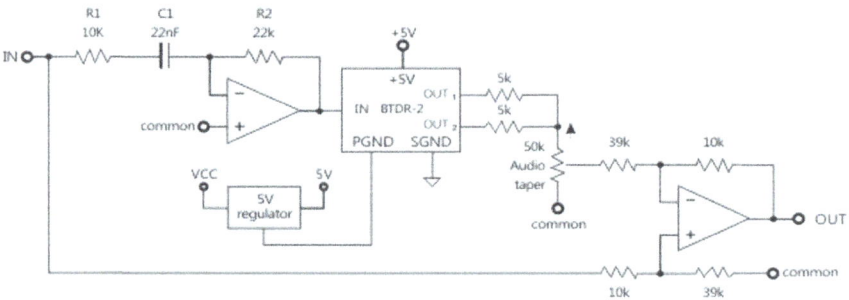

Fig. 9.11: An electric circuit for digital simulation of reverb.

9.4 Time and Pitch Stretching

Let us complete the discussion with a brief explanation of methods that enable tempo and pitch changes in a sound file. We shall focus on the *phase vocoder* method introduced by James L. Flanagan and Roger M. Golden [18] in 1966. First, we will make those effects more precise:

- *Tempo change*: Same pitch, but played at a different tempo,
- *Pitch change*: Same tempo, but transposed pitch.

These operations are very natural when played by a human musician. The former means just to play a piece slower or faster, while the latter means to transpose the piece a number of semitones or keys on the piano. However, this presupposes a symbolic representation of the piece, which then easily enables the performer to map the symbols to different physical parameters. But this symbolic representation is absent in the audio file. Therefore, we need a totally different approach that is independent of the symbolic background.

Such an approach can be given by the phase vocoder method. It uses the Fourier transform of the original sound function. But this is not a straightforward application. In fact, if we take the Fourier amplitude spectrum of a sound wave, showing amplitudes at natural multiples $n \cdot \omega, n \geq 1$, of a fundamental frequency ω, then the transposition to a new frequency ω' produces a *resampling* of the original signal that results in dramatic changes in the sound quality.

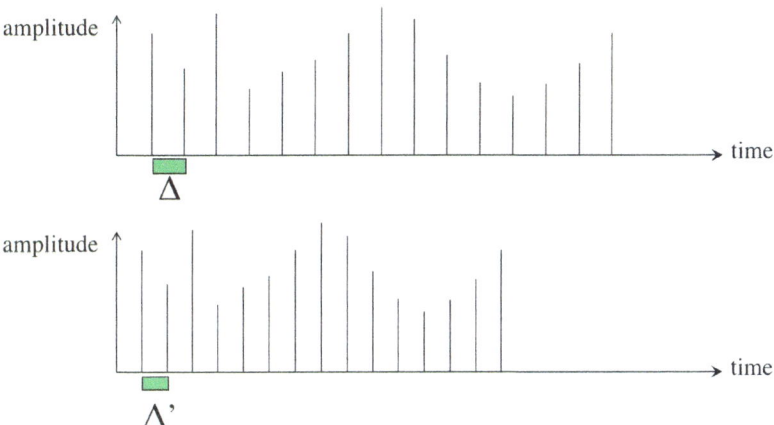

Fig. 9.12: The resampling technique.

In Figure 9.12, the first graph depicts the sound wave, where the horizontal axis represents time, and the vertical axis represents amplitude. The frame period is denoted by Δ. If we shrink the original duration, we produce a new sound wave with a shortened frame period Δ', as shown in the second graph. The new frequency ω' is given by

$$\omega' = \omega \frac{\Delta}{\Delta'}.$$

This frequency transformation dramatically changes the sound's timbre. To change the pitch without distorting the timbre, we will use the following approach:

- Suppose we construct a new sample with *longer* duration and the same pitch. Then we may shrink it back to the original duration by resampling, thus *raising* the pitch.
- Suppose we construct a new sample with *shorter* duration and the same pitch. Then we may stretch it back to the original duration by resampling, thus *lowering* the pitch.

To apply this approach, we must first find a way to change the duration without changing the pitch. Then we may apply resampling to change the pitch and go back to the original duration.

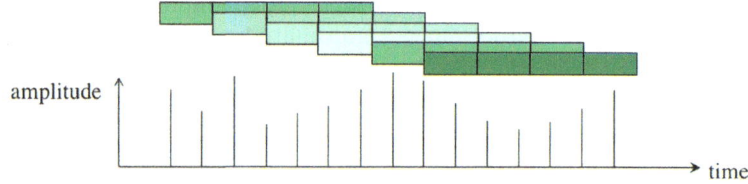

Fig. 9.13: Overlapping frames.

To solve this duration-change problem, we first cover the original sound wave with a sequence of sound frames of equal length. However, in order to grasp their commonalities, we choose overlapping frames. This is typically achieved by covering 75% of the previous overlapped sound frame (see Figure 9.13). Moreover, the frame duration is typically 1/20 second as this corresponds to the fundamental frequency for the finite Fourier, 20 Hz.

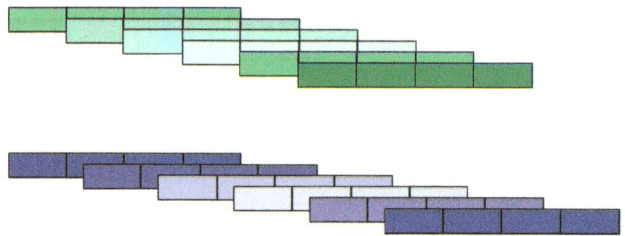

Fig. 9.14: The original overlapping frames (above) and the new overlapping frames (below).

We now work on these frames, process them on the frequency space, and then generate a synthesized sound by adding these new frames with different

9.4 Time and Pitch Stretching

overlapping times, thereby changing the tempo of the overall signal (see Figure 9.14).

A frame is constructed from the original sample by multiplying it by a Hanning window function $H(t)$ (see Figure 9.15). The blank area (located beside the frame) constructed in a Hanning window function (the first graph) has value zero; this zero-value signal is multiplied by the original signal (the second graph). Thus, the multiplication of the signal in the Hanning window function with the original signal results in the desired frame (the third graph) with value zero flanking it smoothly.

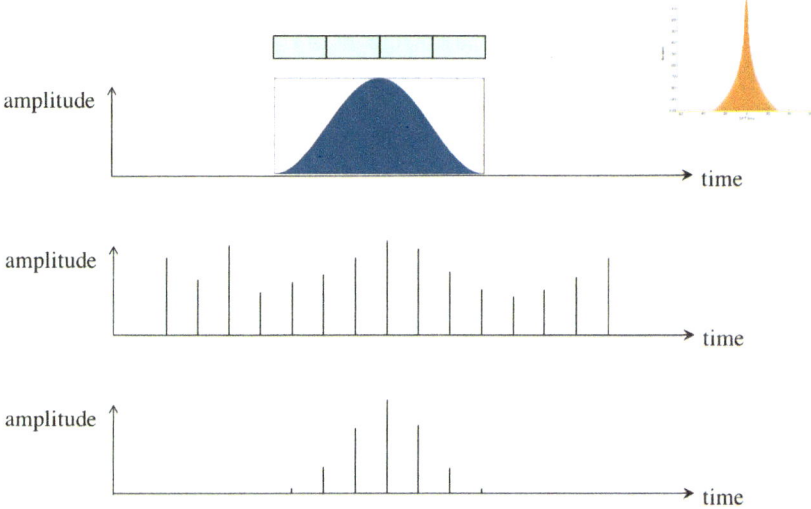

Fig. 9.15: The frame (below) is the product of the Hanning window function (above) and the original signal (middle).

Afterwards, we build the first algorithm through signal analysis. The resulting frame is transformed into its frequency representation via a Fast Fourier Transform (FFT) (see Figure 9.16). The fundamental frequency is $f = 1/$frame duration $= 1/D$. The highest frequency is $f_s = nf$ (see Nyquist's Sampling Theorem p. 77) so we have n frequency intervals from 0 to $f_s(n-1)/n$ Hz.[1] The temporal delay of $1/4$ frame then has $2n/4$ (temporal) samples; this number is called the *analytical hop size*, hop_a. In other words, we have the equation

$$\frac{hop_a}{2f_s} = \frac{2n}{4 \cdot 2nf} = \frac{D}{4} = D_a,$$

where D_a is the analytical *hop time* between successive frames.

---

[1] Notice that n has nothing to do with the original sample frequency of the signal!

104   9 Audio Effects

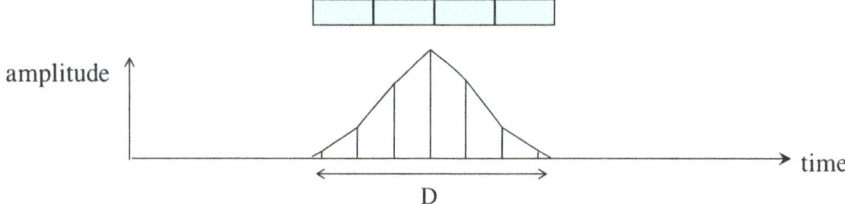

Fig. 9.16: The resulting frame transformed into a frequency representation done via an FFT.

However, at this point, the rearrangement of the frames with a varying hop time, called *synthetic hop time*, causes phase problems (see Figure 9.17). When the frames are shifted into synthetic hop time, a ragged signal will emerge. To find a solution to the phase problem, we then apply a second algorithm in order to produce a smoother signal.

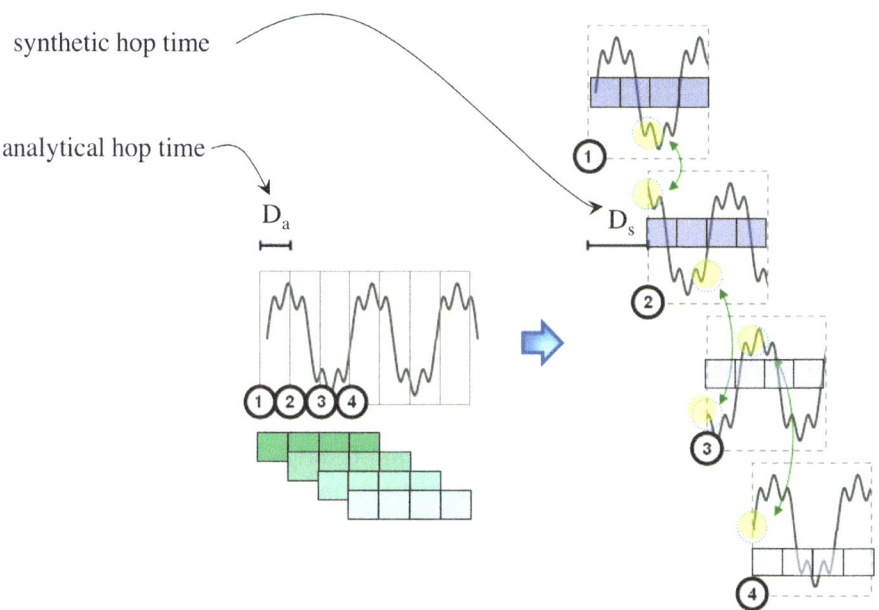

Fig. 9.17: The phase problems caused by the synthetic hop time.

At this step, we must fill in the 'disappeared signal' so that we can produce complete sinusoidal signal components. Suppose we have the original frame signal:

$$sin(2\pi ft),$$

the new phased-frame signal is:

$$sin(2\pi f(t + D_a)) = sin(2\pi ft + \Delta\Phi),$$

with the original frame signal located at frame $(i-1)$, and the phased frame signal located at frame i (see Figure 9.18).

Phase $\Delta\Phi$ can be calculated by manipulating the formula as follows:

$$\Delta\Phi = 2\pi f \cdot D_a, \quad f = \frac{\Delta\Phi}{2\pi D_a},$$

which equals the 'true frequency'.

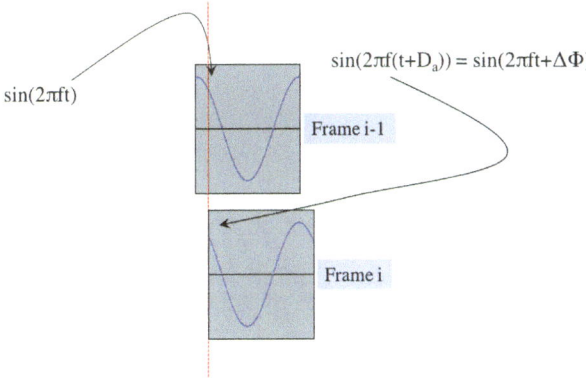

Fig. 9.18: The analysis of the original frame signal (above) and the phased frame signal (below).

The third and last algorithm is a synthesis. We replace D_a with the synthetic frame distance D_s and then set a new phase of frame i:

$$\Delta\Phi_{s,i} = \Delta\Phi_{s,(i-1)} + 2\pi f \cdot D_s,$$

where the $(i-1)$th phase has been calculated by recursion.

Finally, we can correct the complex coefficients of the FFT accordingly and recreate the sound function by inverse FFT (see Figure 9.19). This complex operation produces a smoother signal albeit it does not produce a signal as smooth as the original one.

Fig. 9.19: The complete sinusoidal frames after the final step is undertaken.

Part IV

Musical Instrument Digital Interface (MIDI)

10
Western Notation and Performance

Summary. In previous parts, we discussed ways to record music directly using digital technology. These techniques require us to think of music as precise air pressure variations, and they encode this acoustic data. The advantage of this approach is great accuracy in reproducing sound. However, computers are essential to the process; only a computer can read and write digital sound files. Historically, music was perceived as a series of gestures (see Chapter 2), rather than a string of sound waves. The traditional approach to 'recording' music is therefore to symbolize these gestures instead of the acoustics they create. Such efforts started with the medieval neumes and resulted in modern notation systems that we call *symbolic*.

From neumes to modern Western notation, thinking in symbols has a long tradition in musical thought. We emphasize that musical scores containing such symbols are not simply shorthand for acoustic data. They are far more than acoustical. Misunderstanding them as acoustical entities was virulent in the middle of the twentieth century when composers discovered the precise acoustical structure of sounds. For example, in 1955, in the German school centered around the *Studio für elektronische Musik*, Herbert Eimert tried to downsize all musical notation to the acoustical aspect, arguing that its precision would make all other notations superfluous. As mentioned in the ontology of music (Chapter 2), notes live in a symbolic space, independent of the physical performative realization.[1]

Our goal in writing this chapter is to illustrate the utility of symbolic notations/languages (we use these words interchangeably) in music technology and to identify desirable characteristics of a digital symbolic language. This informs our discussion of MIDI in the following chapters.

$$- \Sigma -$$

---

[1] The difference between symbols and acoustics is described by performance transformations. We discuss these briefly at the end of this chapter.

10.1 Abstraction and Neumes

Musical notation symbols describe gestures, but are not gestures themselves. In other words, symbolic languages are *abstractions* of gestures and are opposed to acoustic data, which has very little abstraction. Notation may be evaluated by its relative degree of abstraction, or how adequately it represents musical performance. For example, cheironomic neumes represent music at a relatively high level of abstraction (see Figure 10.1). This notation, developed in the ninth century for use with Gregorian chants, consists of curves and checkmarks written above text. The earliest forms of these marks indicate only the general direction of changes in pitch (users of this notation likely passed down specific musical information orally). Neumes are purely gestural and provide little to no information about absolute timing or pitch. However, this information is adequate for the purposes of this notation's invention. The point of these symbols is not to reproduce music, but to represent its most relevant and reproducible aspects for performance. We recognize this principle later in modern Western staff notation.

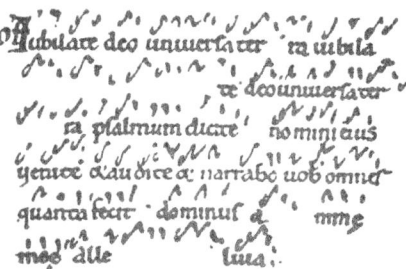

Fig. 10.1: Cheironomic (gestural) neumes above text from Gregorian chant.

Neumes and other abstract symbols are difficult to use in technological applications. Suppose we want to write software that authentically recreates a Gregorian chant from neumes. The original chant started on a particular pitch and took a specific amount of time to sing through. Based on the notation alone, we can merely guess at these values, and different people may interpret the neumes differently; they are both imprecise and ambiguous. These qualities prevent us both from recreating the chant and *a fortiori* from writing software to do so. We learn from this observation that a digital symbolic language must have a lower level of abstraction and fuzziness; it must specify exact values and convey an unambiguous meaning. Furthermore, this language should somehow bridge the gap between gestural and acoustic representation.

10.2 Western Notation and Ambiguity

Before we get into the details, we remark that Western notation is the time-tested result of centuries of experimentation and innovation. Nevertheless, this section explains critically why Western notation is a less-than-ideal language for technological purposes.

In Western notation, pitch is indicated by the vertical placement of note heads on one of five lines or on a space between lines. Tradition dictates that these lines and spaces encode the C major scale (see Figure 10.2). Alterations can be made in single-semitone increments to diatonic notes downward using ♭s, or upward using ♯s. The notation for these alterations is specific in that it indicates both the resulting pitch and the movement (vector) from the original pitch to the altered one. This specificity is useful when working with computers. While this notation represents the typical set of Western intervals effectively, it remains remarkably ambiguous. For example, it does not clarify the basic tuning of the pitch information, whether to use 12-tempered, just, or another tuning. Additionally, the notation does not specify the starting note (whether the note A above middle C should sound at $440\,Hz$, $438\,Hz$, or even at $400\,Hz$). Notes *between* the standard tones are difficult to express. We therefore desire, in a digital symbolic format, a way to specify tuning and pitch unambiguously.

Fig. 10.2: The staff system of Western notation. It is less abstract than neume notation, but remains ambiguous in some respects.

If we want a computer to play music, we must tell it exactly when each note starts and stops. Western notation can be ambiguous in this regard. For example, a quarter note may have any physical duration, depending on the tempo chosen for that performance. The longest instance of John Cage's (in)famous composition *ASLSP* (As Slow As Possible) [8] ever performed will last 639 years (it started on September 5, 2001), see Figure 10.3. Clearly, the physical durations of notes in this piece are extremely long despite their normal appearance.

Tempo,[2] and therefore the time length of each note, is subjective both across multiple performances of a piece and within the piece itself. As a result, the onset time of each note is only fixed relative to bar lines or by its placement in a chord. It is possible to simply measure the tempo continuously throughout

---

[2] Mathematically, tempo encodes the inverse derivative of physical time as a function of symbolic time, see [41, Chapter 6] for details. However, the tempo only encodes the onset of the beginning of a note, not its end, the offset. Offset time can be encoded using the formula offset = onset + duration.

Fig. 10.3: The organ used to perform John Cage's *ASLSP*.

a performance,[3] and therefore reconstruct time. However, this information is not available from the notation alone. In turn, note durations are complicated by the presence of articulation marks, see Figure 10.4, which are completely symbolic and may have different meanings depending on style and context. The fermata is also loosely defined. Essentially, Western notation provides very

Fig. 10.4: Articulation marks make note lengths quite subjective. This is especially true out of context! From left to right, these marks are: *staccato* (separated), *marcato* (marked), *accent* (strong), and *tenuto* (sustained).

limited means of calculating the beginning, end, or duration of any note. A symbolic language for digital applications should therefore define time more precisely.

We have the same situation if we examine loudness information since there is little agreement about what *mezzo forte* (**mf**) should mean in terms of an absolute loudness scale (such as decibels). In Western notation, such symbols are even less specific than pitch symbols. Each dynamical sign tends to represent a range of dynamics and these ranges often overlap. For example, there are situations where *mezzo piano* (**mp**) could have the same loudness as *forte* (**f**) elsewhere.

---

[3] Such information is called the *tempo curve* and may or may not be well defined at any point in the score.

10.2 Western Notation and Ambiguity

Western notation is a symbolic language where sound parameters express huge abstractions of physical reality. Following an insight by Mazzola, and also supported by Theodor W. Adorno's writings about performance [2], musical symbols are "frozen gestures." Performers know that playing a composition from a score means to 'thaw' these frozen gestures until they recover their original 'temperature.' This means that whenever we deal with a Western score, we implicitly think about its gestural kernel. The gestural essence of the score symbols must be recovered completely in a successful performance. The theory dealing with the relationships between these symbols and their physical (instrumental) realization is called *performance theory* and has been an intense research topic in traditional investigations as well as in contemporary research involving quantitative methods and specifically computer-aided models (see [41] for a recent documentation).

To summarize, Western notation is not symbolism for its own sake, but a wrapping and compactification of performative gestures. It is useful both in this regard, and as a foundation for composition. It is also historically relevant that notation was first motivated by theory, it was more than a simple mnemonic device. Despite its musical significance, Western notation does not provide the types of information necessary for software to 'decode' and perform music without significant human involvement. It might be more practical to use a digital notation, such as MIDI, that fills these information gaps.

11

A Short History of MIDI

Summary. We present a concise history of the MIDI technology, an industrial interface for digital communication in music.

$$- \Sigma -$$

MIDI was officially introduced in January 1983 at the NAMM (National Association of Music Merchants) with its "MIDI 1.0 Specification."[1] It replaced the first sketches of a "Universal Synthesizer Interface" from 1981, which had been presented by Dave Smith and Chet Wood at the 70th Convention of the Audio Engineering Society. It is interesting that it was no longer the academic Audio Engineering Society but the thoroughly pragmatic community of music merchants who remained at the wheel. Also, no music-theoretical community was involved in this development, but of course they later complained about MIDI's deficiencies. Standardization of music formats had never been a topic of music theory before mathematical and computational music theory started developing universal standards, see [37, 55, 56, 57].

In the following chapters we describe MIDI communication and the structure of MIDI messages. These messages are exchanged between any two: computers and/or synthesizers. The general functionality of MIDI is shown in Figure 11.1. Making music is a threefold communication: We usually have (1) a score whose frozen gestures are being 'thawed' to (2) gestures, which act on an instrumental interface and thereby produce (3) sound events. The map from score to sound events is the objective of performance research. The gestural interaction with an instrument (process from the left top position to the right top position in Figure 11.1) is where MIDI has its main functionality. The movements of the human limbs (hands for keyboard players) are encoded and then communicated to a synthesizer that produces corresponding sound events. MIDI is therefore essentially a simplified code for human gestures interfacing with an instrument. A second functionality of MIDI is the transformation of

---

[1] The MIDI Specifications are accessible through the MIDI Manufacturer's Association at http://www.midi.org.

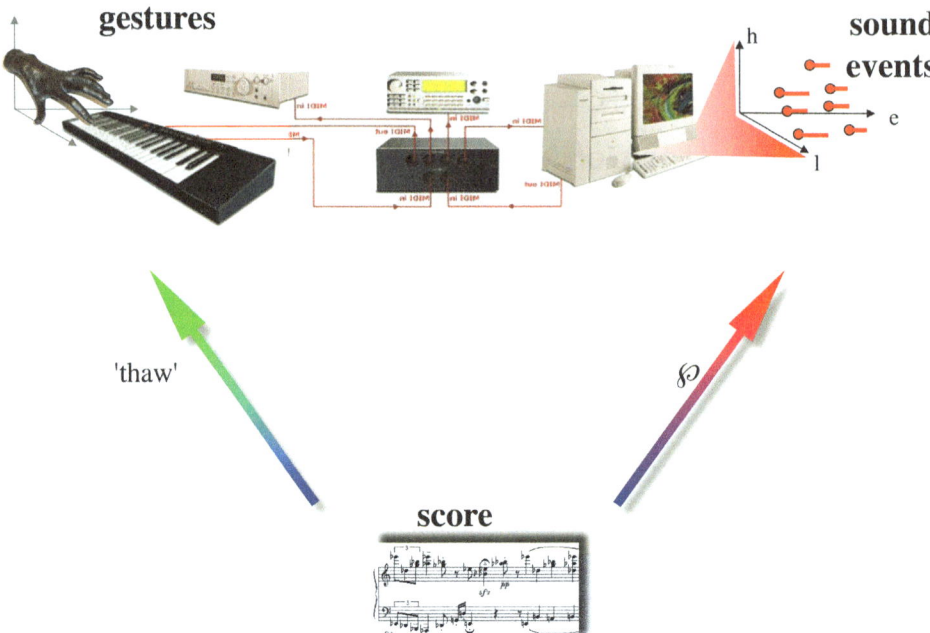

Fig. 11.1: Making music is a threefold communication: We usually have (1) a score whose frozen gestures are being 'thawed' to (2) gestures, which act on an instrumental interface and thereby produce (3) sound events.

instrumental gestures to a *Standard MIDI File* such that the performance can be replayed later. Standard MIDI files can also be transformed into (electronic files for) Western scores, and vice versa. But the essence of MIDI is that it encodes commands of a gestural nature, although in a very simple setup as we shall see in the following discussion. It is not a coincidence that machines that understand the MIDI messages are called *slaves* in MIDI jargon. MIDI just tells an agent (mimicking a musician) to make simple gestures at a defined time and with a specific key and instrument.

It was probably the most ingenious creation in the MIDI code to refrain from abstract symbols and to encode the simple gestural music-playing action instead. The music industry was not interested in higher symbols but in a communication code that would help musicians play electronic instruments when performing onstage or in a studio. We see here an interesting schism between codes and thoughts in theory versus performance.

12
MIDI Networks

Summary. In this chapter, we discuss how MIDI devices, ports, and connectors work. The basic MIDI devices include sequencers, synthesizers, samplers, and controllers. Most such devices have three MIDI ports, namely **In** port, **Out** port, and **Thru** port. We can build networks by connecting devices, ports and connectors.

$$-\Sigma-$$

12.1 Devices

Controller: The MIDI controller controls other devices. It does not make a sound, but it lets the player play the 'notes' and send the corresponding message to other devices. The most common controller is the keyboard controller. There are also other controllers, such as drum, guitar, and wind controllers. The variety of controllers allows musicians to use different instruments in the MIDI system.

Sequencer: The sequencer is the core of the MIDI system, and it can either be in the form of hardware or software. Similarly to the multitrack recorder in the recording studio, the sequencer can record the music. However, instead of recording and playing the sound, the sequencer records and plays the messages. The messages represent what instrument to use, the dynamic and tempo of the piece, etc. The messages can be recorded track by track into the sequencer from MIDI devices. After the sequencer records what has been played—whence its name—the tracks can be edited later. Usually, individual sequencers have about 9-16 tracks, while the software can have more than 200 depending on the hardware's memory.

Synthesizer: Synthesizers are electronic instruments that produce electrical signals. Synthesizers can generate sound by mimicking different instruments and also create new sounds based on the existing ones. This enables the producer to be more creative in producing music in different forms. Synthesizers use *oscillators* to generate sound from electronic signals (waveform), *filters* to filter out the unwanted frequencies, and *amplifiers* to control the sound volume.

Sampler: Samplers are similar to synthesizers but instead of generating a new sound, they record sound from the real world as samples, and that is why they are called samplers. After recording, a sampler converts the sound into electrical signals and maps each sample onto a key on the keyboard. Samples can also be pitch-shifted to build chords or scales, etc. However, the pitches can be made less 'natural' by the shifting process, so sometimes a sound is recorded multiple times at lower and higher pitches.

12.2 Ports and Connectors

MIDI communication is done via MIDI cables. They accept MIDI messages at an **Out** port and pipe them either to an **In** port, where a machine uses the message, or to a **Thru** port, which makes the message available to an **Out** port of the same machine for further messaging to other machines. The three port types are shown in the lower center of Figure 12.1. MIDI messages are Bit sequences sent at a frequency of 31,250 Bits per second (= Bauds) with a current of 5 mA. Equivalently, this is 1/32 of a MegaBaud. This means that the transfer of one Bit needs 32 μsec. Originally, such transfer was strictly serial (one Bit after another), but nowadays there is also parallel data transfer.

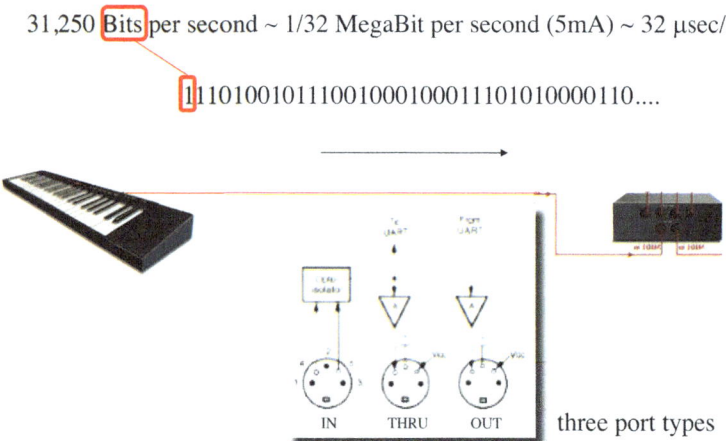 three port types

Fig. 12.1: MIDI messages are Bit sequences, and the ports connecting MIDI devices have three types: IN, THRU, OUT.

MIDI is semiotic, and the MIDI Specification connects the expressions given by Bit sequences to their symbolic signification (see Figure 12.2). But we should be precise here: The contents of MIDI messages are not directly musical.

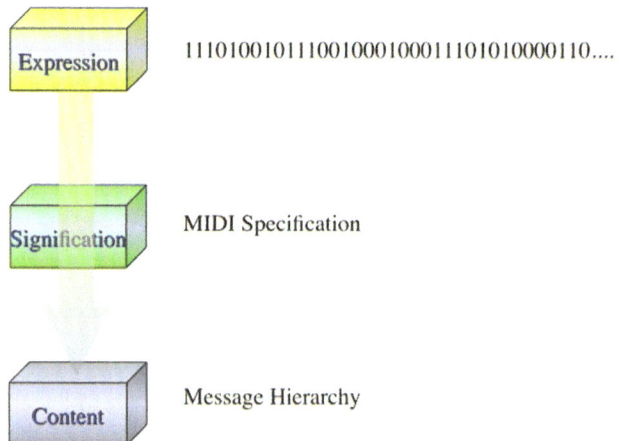

Fig. 12.2: The semiotic role of a MIDI Specification.

They just decipher Bit sequences and yield information that a synthesizer can understand, but it is the synthesizer's semiotic (!) that must transform the message contents to musically meaningful information. We will clarify this when looking at the message structure.

13
Messages

Summary. This chapter introduces the basic idea of MIDI messages and explains different forms of MIDI messages. MIDI messages are the "orders" sent through devices and the "keys" for the communication between different devices. There are in total seven basic MIDI message types (Note Off, Note On, Polyphonic Key Pressure, Control Change, Program Change, Channel Pressure, and Pitch Bend). Knowing what each type of message means and does can help us understand more about how MIDI works.

$$- \Sigma -$$

13.1 Anatomy

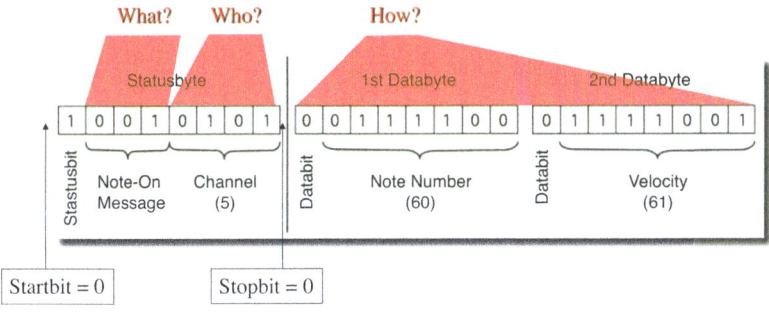

Fig. 13.1: The MIDI message anatomy.

A MIDI message consists of a sequence of words, each having a length of 10 Bits (see Figure 13.1). These words are delimited by a zero startbit and a zero stopbit, and it takes 320 μsec per word, which means roughly a third of a millisecond. MIDI messages are sequences of a few words with a fixed content anatomy. The first word is the *status* word, and the status character of the word is initiated by the first Bit: the *statusbit* one.

The word encodes two information units: (1) what is being played, encoded in three Bits; and (2) who is playing, encoded in four Bits. We will come back to these contents later. The second portion of the message is composed of a number of words. They are all—on the Byte (= 8 Bits) inside their start- and stopbits—initiated by the *databit* zero. These words encode information about how the actions, as defined by the status word, are played.

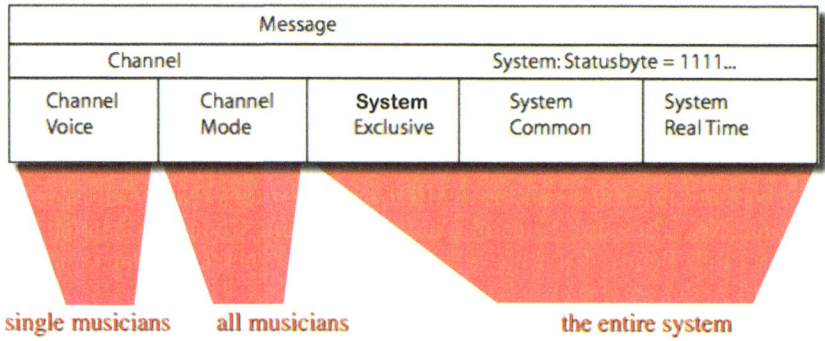

Fig. 13.2: The MIDI message types.

Let us first consider the example shown in Figure 13.1. Here the first three Bits 001 encode the action of pressing down a key (Note On, the note's onset action). The following four Bits 0101 tell us which 'musician,' called *channel* in the MIDI code, is playing the Note On action; here it is channel 5, written in its binary representation 0101. With this information, the data words tell the musician which key to press down: The seven Bits in the first data word, namely 0111100, correspond to 60, the middle C on the piano. The second data word shows the seven-Bit binary number 1111001, corresponding to 61, and meaning "velocity." Here we see the gestural approach: "Velocity" means the velocity with which a keyboard player hits key number 60, and this is a parameter for loudness. The data words can specify any number in the interval $0-127$, so the velocity value 61 is a middle loudness. The total time for such a message is three times 320 μsec, roughly one millisecond. This is not very short, and it can in fact happen that complex musical information flow can be heard in its serial structure. Recall that every single command has to be sent serially. For example, a chord is sent as a sequence of its single notes.

As announced above, these numbers are still quite symbolic. The pitch associated with a key number is not given in the MIDI code, so it is up to the synthesizer to interpret a key number in terms of pitch. And velocity must also be specified by the synthesizer in order to become a loudness value. This is a typical situation of a connotative semiotic system (meaning that signs are expressions of signs of a 'higher' semiotic system): The MIDI Specification only decodes the Bit sequence, and then this result is taken as an expressive unit pointing to the acoustical contents via the synthesizer's signification engine.

hex	Statusbyte	Databyte	Message	Description
$8n	1000nnnn	0kkkkkkk	Note Off	kkkkkkk: Key #
		0vvvvvvv		vvvvvvv: Velocity
$9n	1001nnnn	0kkkkkkk	Note On	kkkkkkk: Key #
		0vvvvvvv		vvvvvvv: Velocity
$An	1010nnnn	0kkkkkkk	Polyphonic Key	kkkkkkk: Key #
		0vvvvvvv	Pressure	vvvvvvv: Velocity
$Bn	1011nnnn	0ccccccc	Control Change	ccccccc: Controller #
		0vvvvvvv		vvvvvvv: Value
$Cn	1100nnnn	0ppppppp	Program Change	ppppppp: Program #
$Dn	1101nnnn	0vvvvvvv	Channel Pressure	vvvvvvv: Pressure
$En	1110nnnn	01111111	Pitch Bend	1111111: LSB
		0mmmmmmm		mmmmmmm: MSB

Fig. 13.3: Some channel voice messages.

13.2 Hierarchy

The messages are divided into different categories (which are then encoded in the statusbyte). These categories are shown in Figure 13.2. There are two big categories of messages: channel and system messages. The former relate to single voices or their collaborative action, whereas the latter pertain to the entire system and are encoded by the statusbyte starting with 111 after the initial status index 1. Channel messages of voice type relate to single musicians (channels), while channel mode messages relate to the channels' collaboration. Figure 13.3 shows some channel voice messages. You recognize the 1 as the initial Bit of the statusbyte. The action type is 1000 for Note Off; the databyte starts with the mandatory zero and then has the 128 possible values indicated by $kkkkkkk$. The two databytes define Key number and Velocity, as in our example above. Program Change, for example, describes the instrument that is going to be played. This is a typical situation where the meaning depends upon the synthesizer's settings. For the so-called General MIDI standard, these numbers have a fixed meaning, 0 being reserved for acoustic grand piano and 127 being assigned to gunshot.

The system messages are of three types: *System Exclusive* for messages that are reserved for specific information used by the industrial producer, such as Yamaha, Casio, etc.; *System Common* messages for general system information, such as the time position in an ongoing piece; and *Real Time* messages for time information (e.g., the timing clock that produces the ticks; see Chapter 14 about ticks).

14

Standard MIDI Files

Summary. In this chapter we discuss the time information encoded in Standard MIDI Files. We begin by explaining the importance of storing time information in Standard MIDI Files, then describe the treatment of time in MIDI, and finally provide details of how time information is defined within Standard MIDI Files.

$$- \Sigma -$$

14.1 Time

When synthesizers[1] create and interpret MIDI messages and 'project' them into acoustical content, they decode Bit sequences in real time. But the time position of such a message is not given within the MIDI message as such. Therefore, when the MIDI messages are compiled into a file, time information is required.

For storage purposes, MIDI has a special treatment of time. It is not a physical time format, but counts time in multiples of *ticks*. The value of a tick is variable. Usually a tick means a 24th of a quarter note. But the physical duration of a tick must be defined on a file's header. In Standard MIDI Files, this header contains meta-information that informs us about the tempo that relates ticks to physical durations.

14.2 Standard MIDI Files

Standard MIDI Files save MIDI information, including data about time, in order to play a recording back on a MIDI device. The *Standard MIDI File Formats*[2] define the syntax of such files. Let us look at a typical Standard

---

[1] MIDI controllers, keyboards, computers, smartphones, etc.
[2] There are various kinds of formats for more or less complex music formats.

Standard MIDI File

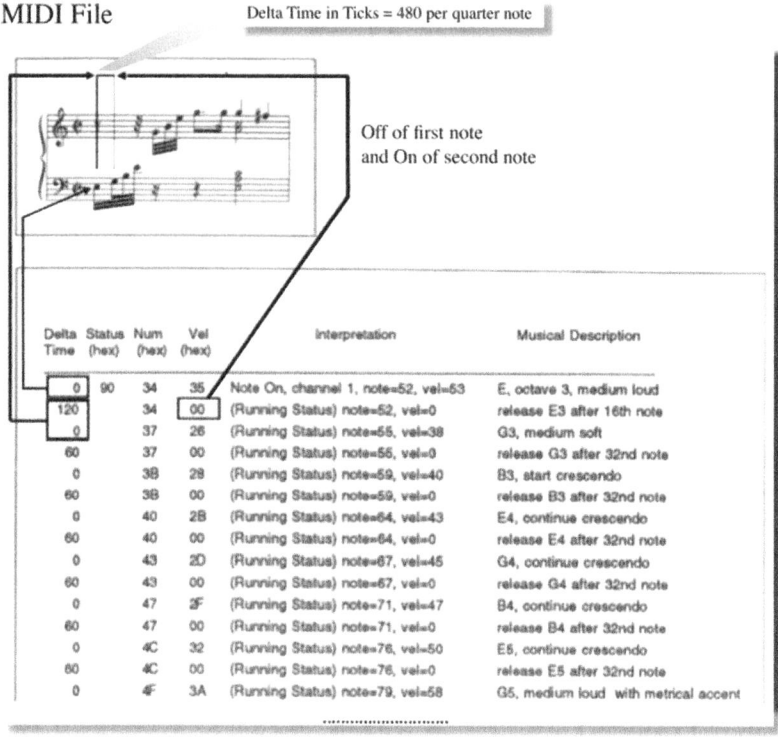

Fig. 14.1: A typical Standard MIDI File showing how this syntax looks in a concrete piece of music.

MIDI File to see what this syntax looks like in a concrete piece of music (see Figure 14.1).

A Standard MIDI File is encoded as a numerical table. In the first column of the table we see the delta time, which indicates in multiples of ticks the time delay between successive messages. The second column indicates the status byte in hexadecimal (base 16) numbers. The action of the first row is Note On, encoded by 1001, and the player is 0000 (first channel 0), yielding the hexadecimal pair 90.[3] The next number is the hexadecimal key number 34, i.e., decimal key number $52 = 3 \times 16 + 4$. This corresponds to the note E3, the first note in the score at the top of the figure. Then we have the hexadecimal velocity value 35, i.e., the decimal MIDI velocity 53, medium loud. The next event is 120 ticks later and instructs the player to play the end of the first note. Here we have a very economical convention, namely the "running status" command, which means not to change the status. This means that we again play that E3 but now with zero velocity! We play the note again but mute it.

---

[3] Binary 10010000 is decimal 144, and this is 9×16, i.e., hexadecimal 90.

This is a creative use of the Note On message! When done with this note, we start the second note, G3, without any time delay, i.e., the delta time is zero after the end of the first note, and so on.

Part V

Software Environments

15
Denotators

Summary. Denotators are a universal data format for musical objects. They generalize the Internet-related XML language and were introduced for music theory, but also for data management in music software for analysis, composition, and performance.

$$- \Sigma -$$

Fig. 15.1: Jean le Rond d'Alembert and Denis Diderot, the fathers of the famous French *Encyclopédie* first published in 1751.

Denotators were introduced to music informatics in 1994 [36] as a data format of the RUBATO® software to cover representations of all possible musical objects.

In search of a powerful data format for music, Mazzola and his collaborators started with an approach that resembled the well-known database management systems (DBS). They tried to be flexible on the parameters that represent musical objects. They set up lists of sufficient length to cover onset, duration, loudness, glissando, crescendo, instrumental voice, etc., so every such object would live in a big space, and most objects would only need a fraction of the available parameters. For example, a pause would have vanishing loudness, a fictitious pitch, and only onset and duration and voice as reasonable parameters. On the one hand, this looked artificial—why wouldn't we just take the parameters that were really used? On the other hand, they were also questioned by ethnomusicologists who made it clear that one would never achieve a universal representation; every day they find new instruments, techniques, and parameters that had not been considered in the past.

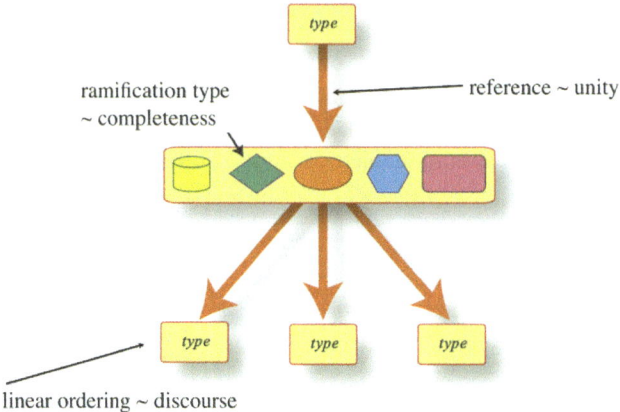

Fig. 15.2: The architecture of conceptual construction as suggested by the French encyclopedia.

The creative solution of this dilemma was the negation of the requirement of a single top space. Mazzola and his collaborators needed a data format that could add new object types at any time without having to revise the entire format, as would be the case in database management systems. It became clear that this meant creating a multiplicity of spaces, one for each object type. This would however be the end of universality since no systematic construction would be at hand. They found the solution from two sources: (1) The semiotic theory of French encyclopedists Jean le Rond d'Alembert and Denis Diderot, the authors of the famous *Encyclopédie*, published in 1751, see Figure 15.1, and (2) The universal constructions in modern mathematics—limits, colimits, and power objects—as universally present in a topos, the most powerful single mathematical structure of the last 50 years, in fact the basic structure of geometry and logic, of theoretical physics, and of music theory.

The semiotics of the French encyclopedists is based upon three principles:

- *unité* (unity) - grammar of synthetic discourse - philosophy

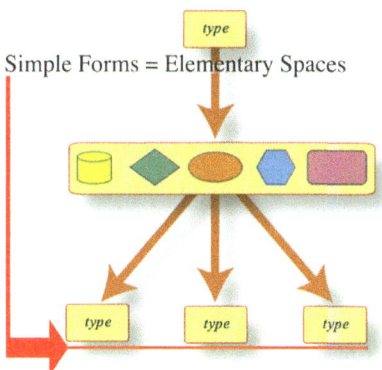

Fig. 15.3: The recursive ramification principle.

- *intégralité* (completeness) - all facts - dictionary
- *discours* (discourse) - encyclopedic ordering - representation

It became clear from the mathematical source that concepts are built according to the French principles by a quite explicit methodology, which means that the multiplicity of spaces postulated when forgetting about the idea of a unique space could be built following a unifying method. The three French principles would offer an architecture of conceptual construction, illustrated in Figure 15.2. The idea is that every concept type refers to other concepts. This is what we understand as being unity. Next, references are given by a comprehensive ramification variety, the ways to refer to previous concepts (see Figure 15.3). This means completeness: There is no other way to build concepts. And third, these types can be ordered in some linear ordering and thereby shape the discursive display of conceptual architectures.

These principles now apply to define concretely what are our concept architectures (see Figure 15.4). According to the encyclopedic approach, the basic building principles are recurrence, ramification, and linear ordering. Let us first start with what is being described. We have to define concepts that are of a specific type (its 'house'). We shall no longer suppose a big universal house, but, like snails carrying around their own houses, for each concept a special house. In geometric terms (and also referring to the object-oriented programming paradigm), we shall view concepts as being points in a given geometric space. We call these points *denotators* and their spaces *forms*, similar to the Aristotelian distinction between form and substance, denotators being substance.

A form must have a name (a sequence of ASCII symbols, see also Section 20.2.1 for more details about ASCII), the form_name. Also a denotator must have a name, the denotator_name. This is necessary because not only values count but also their signification. For example, a real number might be an onset or a duration. This is taken care of when specifying the form name as being

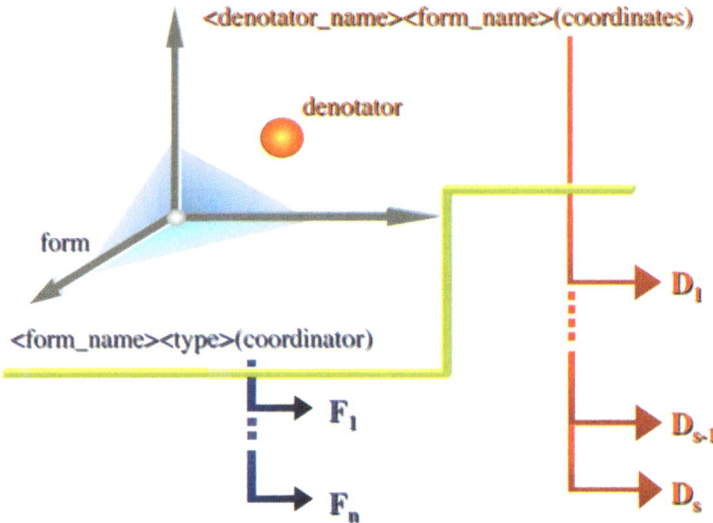

Fig. 15.4: Our concept architecture of denotators and forms.

"Onset" or "Duration." Next, the ramification *type*, i.e., the way the form refers to other, already given forms, must be indicated. We shall see in a moment what types are available. The type being given, the collection of referred forms is called the *coordinator*. It will in general be a sequence $F_1, \ldots F_n$ of forms. The denotator shows its form in the entry for form_name. According to this form, the denotator has a collection of coordinates, which are also denotators, usually a sequence $D_1 \ldots D_s$ of denotators.

In our form architecture, there is a recursive basis, namely the *simple forms*. They are those spaces that contain purely mathematical information. In this case, the coordinator is a mathematical space A. Let us look at a concrete space (see Figure 15.5). The form name is "Loudness," the type is "Simple," the coordinator is the set $A = STRG$ of strings from a given alphabet, usually ASCII. A denotator may have the name "mezzoforte," the form's name being "Loudness," and the coordinates are a string of letters from $STRG$, in our case *mf*, the symbol used in scores for a medium loudness. We may vary the mathematical space; here are some simple first examples:

- form = <HiHat-State><Simple><Boole>,
  denotator = <openHiHat><HiHat-State><YES>
- form = <Pitch><Simple><$\mathbb{Z}$>,
  denotator = <thisPitch><Pitch><58> (b-flat)
- form = <Onset><Simple><$\mathbb{R}$>,
  denotator = <myOnset><Onset><11.25>
- form = <Eulerspace><Simple><$\mathbb{Q}^3$>,
  denotator = <myEulerpoint><Eulerspace><(3/4,-5/6,2/1)>

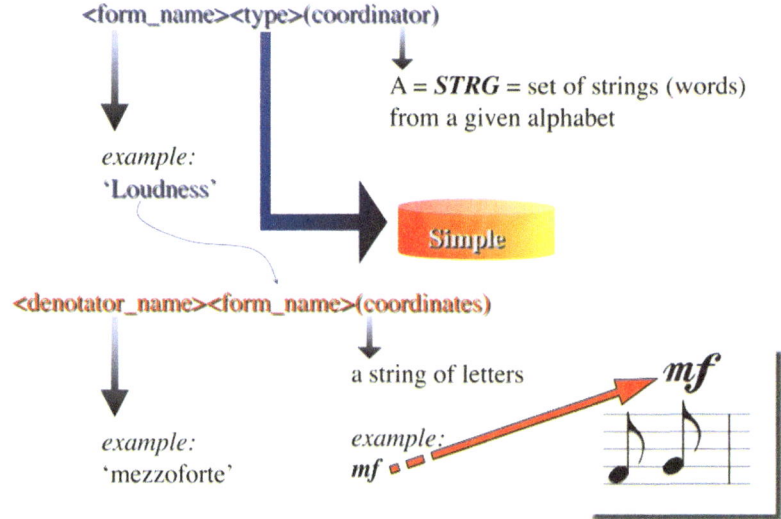

Fig. 15.5: Simple forms referring to $STRG$

The most general simple space scheme is to take any module[1] M over a commutative ring R and select the coordinates of denotators as being elements $c \in M$. Examples are pitch classes with module $\mathbb{Z}_{12}$ or a module $M = \mathbb{Z} \times \mathbb{Z}_{365} \times \mathbb{Z}_{24} \times \mathbb{Z}_{60} \times \mathbb{Z}_{60} \times \mathbb{Z}_{28}$ for time in years:days:hours:minutes:seconds:frames. Why only modules? This has no deeper reason; it is just the straightforward mathematical basis from the examples of classical theories and practice. But one is now also envisaging topological contexts, in particular for mathematical gesture theory [40].

The recursive construction is the part that relies on the universal operations covered by the topos structure. We call such forms *compound*, and it is here where the ramification must be specified: How do we refer to given forms? The first compound type is called *limit*, and it generalizes the Cartesian product known from set theory. An example is shown in Figure 15.6. This example has form name "Interval." Its coordinator is a diagram of already given forms; here we have n forms $F_1, \ldots F_i, \ldots F_n$. These forms are connected by arrows that designate transformations between the forms' denotators. These transformations generalize the affine transformations we have in the case of modules of simple forms. In our example we have three forms: two named "Note," and one named "Onset." The example shows two transformations from the Note form to the Onset form, namely projecting a note denotator to its onset denotator. A denotator here called "myInterval" contains two Note denotators plus one

---

[1] Modules are generalized vector spaces. The difference to the latter is that the scalars in modules need not be fields, such as real numbers, but can be commutative rings, such as the integers or polynomials.

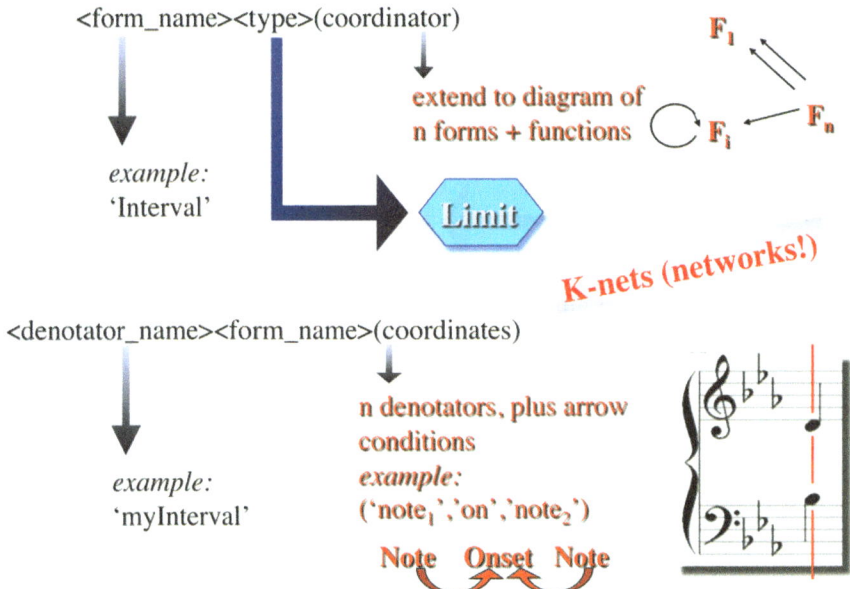

Fig. 15.6: Limit forms.

Onset denotator, and the transformations guarantee that the onset denotator is identical to the onset of both notes. This simply means that we have a couple of notes at the same onset, i.e., a simultaneous interval. Limit denotators generalize what in American music theory following David Lewin's transformational ideas has been called a Klumpenhouwer network [39].

The second universal construction of compound forms relies on the "dual" of limits, namely *colimits*, which generalize the union of sets. In our example in Figure 15.7, we describe a colimit form defined by the same diagram of forms as above for the colimit construction. The denotators are the elements of the union of the denotators of all forms $F_1, \ldots F_i, \ldots F_n$ modulo the identification of any two such denotators that are connected by a chain of the transformations in the diagram. In the example here, we take just one form $F_1 = $ **Chord** parameterizing chords of pitch classes (see the next construction for details of this form), with one transformation, the transposition by n semitones. The colimit consists of the chord classes modulo n-fold transposition.

A third universal construction of compound forms is called *powerset*. Its coordinator is a single form F, and the denotators are just sets of denotators of form F. In our example in Figure 15.8, the "Chord" form is defined, and its denotators are sets of denotators of form $F = $ **PitchClass**.

This fourfold typology: simple, limit, colimit, and powerset is everything one needs to perform any known conceptualization in mathematics and also in music theory. The system has also been used to model *GIS* (Geographic

15 Denotators

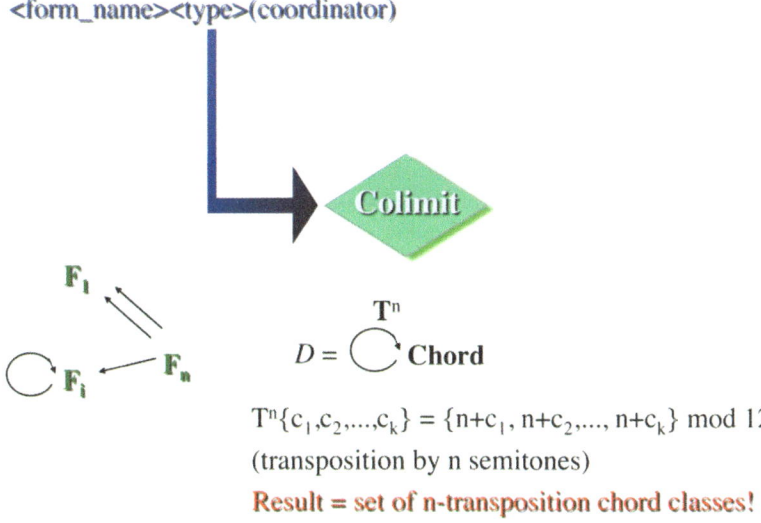

Fig. 15.7: Colimit forms.

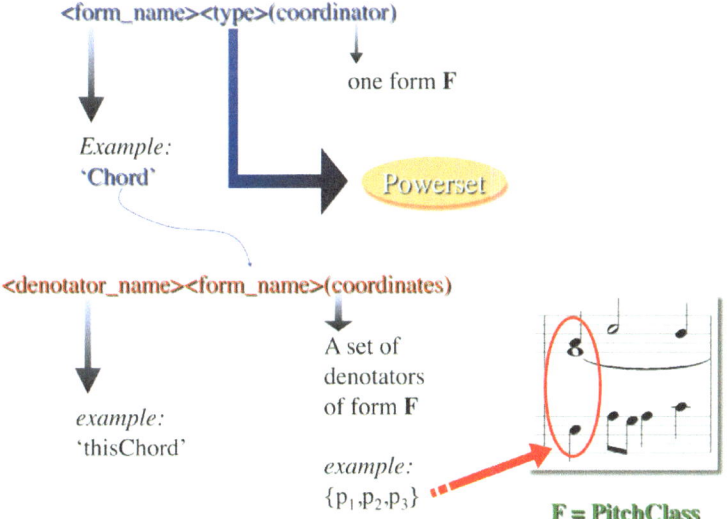

Fig. 15.8: Powerset forms.

Information Systems). The creative point here is the very small construction methodology for a variety of spaces. And we should add that these techniques have also been implemented in a Java-based program RUBATO® COMPOSER [45], where forms and denotators can be constructed on the fly, whenever one needs a new form or denotator.

However, there is a detail that is astonishing when such a creative extension of the space concept is at hand: We did not exclude circularity, which means that in the reference tree of a form, it may happen that the same form reappears. At first, this seems pathological, but it turns out that an important set of forms use effectively this circularity, e.g., forms for Fourier analysis or forms for FM sound representation. More than this: It turns out that essentially all interesting formats for notes are circular. For example, notes that have satellites for ornamentation or Schenker analysis are circular, see [45]. Therefore the creative act not only solves the original problem of the multiplicity of spaces, but it even enables space constructions that were not anticipated by the classical understanding of what a space can be.

16
Rubato

Summary. In this chapter, we introduce the RUBATO® COMPOSER software and provide basic information for the subsequent chapter about the Big-Bang rubette, an interesting RUBATO® component.

$$-\Sigma-$$

16.1 Introduction

Following the invention of the digital computer, computation has become essential to an increasing number of fields. Musicians have also sought programs to help create and analyze music. For example, in the past, sonatas were only composed and analyzed using simple tools such as pencils, graph paper, rulers, etc. Computer software was developed to enhance the composition process. RUBATO® is an example of such software.

The goal of the RUBATO® software was to bring accessible mathematical and computational tools to composers. The software is based on module theory, an essential part of mathematical music theory. The idea of modules generalizes group theory, and allows us to represent most musical objects, including notes and chords. We can even represent transposition, inversion, retrograde, and the entire theory of counterpoint through modules. The denotator concept that we introduced in the last chapter is based on module theory and is used throughout the RUBATO® software.

16.2 Rubettes

RUBATO® provides tools to manipulate musical parameters. These are accessible on the nodes of a data flow network. Regarding data flow networks, it is essential to specify what constitutes a node. In RUBATO® such a component is called a *rubette*. Rubettes are implemented based on the idea of computational theories.[1] The data flow approach in RUBATO® translates mathematical objects into denotators, which flow through the network of computational units, or rubettes.

Take an example from the area of number theory. If integers are objects, one possible computational unit would be to calculate an integer as the product of two factors. The input of the unit would be an integer and the output its factors. The output of such factorizations can then be the input to another computational unit which calculates the greatest common divisor. But a rubette is usually not merely a computational primitive such as the simple addition of numbers; it is often programmed to perform quite sophisticated calculations.

Rubettes may also provide direct manipulation through two graphical interfaces presented to the user as a view where data and properties that change the behavior of the rubette can be rendered in an audio-visual manner. Besides the input and output ports, a rubette also has display properties to access the processual parameters, as well as a view dialog for visual display and interaction, see Figure 16.1.

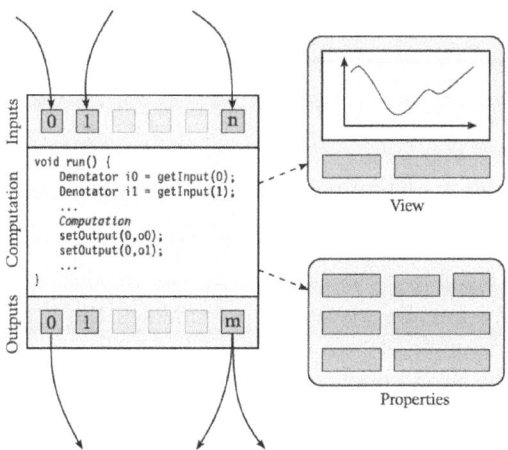

Fig. 16.1: The components of a rubette: input and output ports, the rubette's program, and the two graphical interfaces: View and Properties.

The RUBATO® system is a platform for two types of users: the developer who implements rubettes and the designer who constructs the network. Examples of rubettes will be discussed in the next chapter.

---

[1] A computational theory is a theory that contains a significant computational part, or, simply stated, a theory that can be implemented as a computer program.

16.3 The Software Architecture

The architecture of the software contains four layers, I, II, III, and IV, which are shown in Figure 16.2. Layer I is the foundation of the software, which is based on

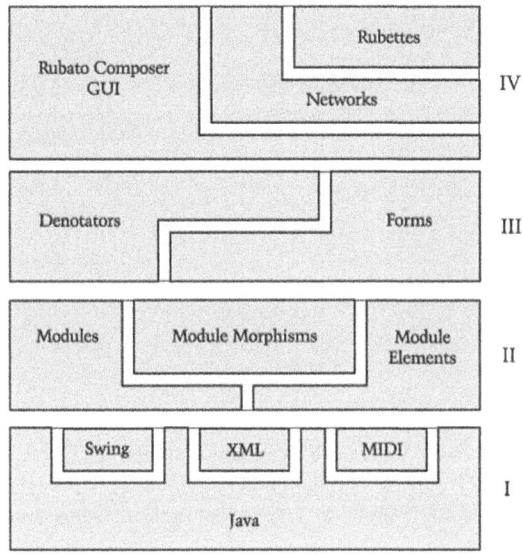

Fig. 16.2: An overall structure of RUBATO® Composer.

the Java environment. The Java classes include classes for MIDI management, XML classes for saving and loading RUBATO® files, and Swing classes for the graphical user interface management. There are also some extensions such as XML Reader and Writer classes and other utilities. This layer also includes the implementation of rational and complex numbers.

Layer II includes the implementation of all the mathematical classes on which the system is based. These include mathematical modules, module elements, and module morphisms.

Layer III is the core of RUBATO®. It contains the implementation of forms and denotators. Like modules and elements, forms and denotators are instances of classes in a parallel hierarchy. They consist of implementations for each type of form: Simple, Limit, Colimit, Power, and an additional type List. Forms and denotators support any number of generic operations.

Layer IV is the application layer. Unlike the layer III, which was mainly for programmers who are familiar with Java, this layer provides the user with an intuitive graphical user interface (GUI). This layer implements rubettes as well as the GUI for interaction. A network GUI is shown in Figure 16.3.

16 Rubato

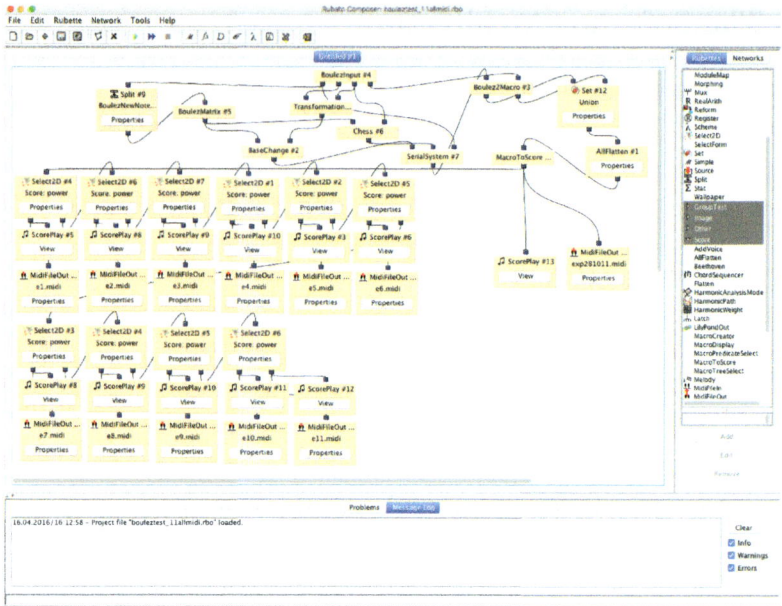

Fig. 16.3: A network of rubettes in RUBATO®.

17

The BigBang Rubette

Summary. The BigBang rubette is a gestural music visualization and composition tool that was developed with the goal of reducing the distances between the user, the mathematical framework, and the musical result. In its early stages, described for instance in [51, 52], it enabled the definition, manipulation, and transformation of *Score* denotators using an intuitive visual and gestural interface. Later, it was generalized for transformation-theoretical paradigms based on the ontological dimension of embodiment, consisting of facts, processes, and gestures, and the communication between these levels [53].

$$- \Sigma -$$

BigBang is a regular RUBATO® COMPOSER rubette, as introduced in the previous chapter, with a variable number of view windows where users can easily create denotators by drawing on the screen, visualize them from arbitrary perspectives, and transform them in a gestural way. On a higher level, they can interact with a visualization of their compositional or improvisational process, and even gesturalize the entire process in various ways. BigBang simply has one input and one output and now accepts almost any type of denotator as an input, to be visualized and interacted with. Figure 17.1 shows the rubette, incorporated in a RUBATO® network, and one of the rubette's view windows including the so-called facts and graph views.

All this is made possible through an architecture based on the three levels of embodiment. The *facts view*, the large area on the right, visualizes the musical objects or facts, which are simply denotators in their coordinate space. It allows for different views of these objects, which are configurable using the grid of checkboxes on the right. The smaller area on the left, the *process view*, visualizes the graph of the creation process of the music, in a similar way to graphs in transformational theory. Each arrow corresponds to an operation or transformation, while each node corresponds to a state of the composition. The graph can also be interacted with and used as a compositional or improvisational tool itself. Finally, the *gestures* are visualized in the facts view, whenever the represented musical objects are interacted with. Any operation or trans-

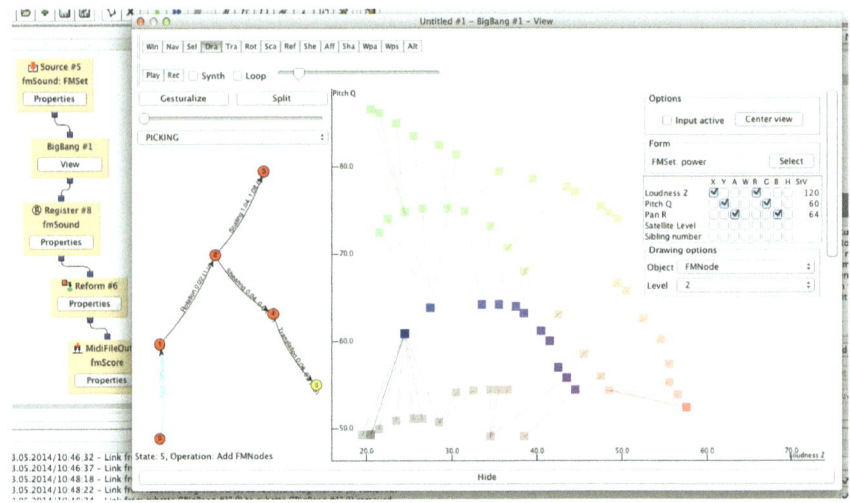

Fig. 17.1: A network including the BigBang rubette and its view next to it.

formation performed in BigBang is immediately and continuously sonified and visualized and can be reconstructed at any later stage.

Apart from this functionality, BigBang behaves just like any other rubette: Whenever the user presses on RUBATO®'s run button, BigBang accepts a denotator, either adding it to the one already present or replacing it (depending on the user's settings), and sends its previous denotator to the next rubettes in the network. The rubette can be duplicated, which copies the graph in the process view along with any denotators created as part of the process. This way, users can include a BigBang with a defined process in other parts of the network, or other networks, and feed them with different inputs, while the process remains the same. Finally, like any rubette, BigBang can be saved along with the network, which again saves processes and corresponding facts.

A more detailed description of the architecture and implementation can be found in [44]. Examples of gestural improvisation with the BigBang rubette are available on [5].

18
Max®

Summary. In this chapter, we will discuss the visual programming language Max®. We first introduce the software, then explain its short history, explore its software environment, and elaborate some technical details within the software.

$$-\Sigma-$$

18.1 Introduction

Max®, sometimes called Max/MSP, is a visual programming language that has a graphically interactive interface for music, audio processing, and multimedia [10, p. XI]. Max® uses a graphical interface to display each basic functional unit on the screen as a small rectangle called an *object* [27, p. 2]. Throughout the world, Max® is used by performers, composers, sound designers, visual artists, multimedia artists, educators, game developers, and IT programmers.

To specify processes in Max®, users build programs by creating objects and connecting them with virtual input and output cables [10, p. XI]. Input cables are located on the top of objects (known as *inlets*) whereas output cables are located on the bottom of objects (known as *outlets*). A document in which users create objects is called a *patcher* or a *patch* and the virtual cables are called *patch cords* (see Figure 18.1). Those patches may perform calculations, synthesize or process sounds, or render visuals, and users may also design a specific graphical user interface (GUI). Because of this customization, Max® works similarly to modular synthesizers where each module[1] manages a unique task or function, connects with other modules, and exchanges data with the connected modules [10, p. XII].

---
[1] Attention, here the term "module" is meant in the sense of a functional unit, not as the mathematical structure that we presented for denotators and forms in RUBATO®.

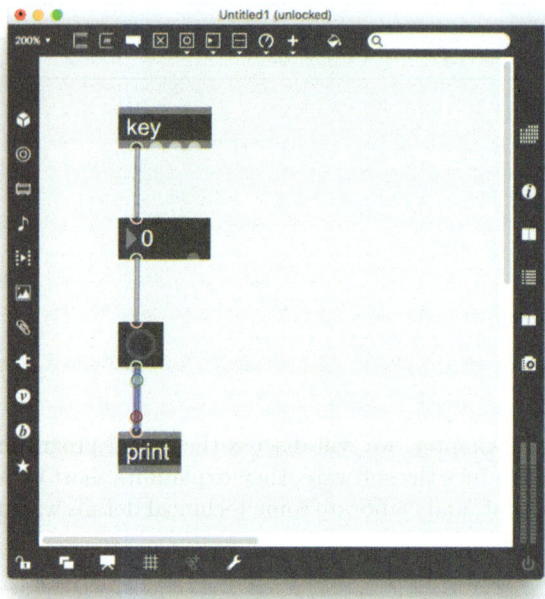

Fig. 18.1: A patch showing four objects (from top to bottom), *key*, *number*, *button*, and *print* connected with patch cords.

18.2 Short History

The history of Max® starts in 1987, at IRCAM (Institut de Recherche et Coordination Acoustique/Musique), in Paris. There, Miller Puckette with the help of Philippe Manoury started working on a new software that would be able to follow and complement musicians during their performance, just like a human musician would. In Puckette's words [9, p. 183]: "My idea was to establish a connection between a soloist and a virtual orchestra, but the orchestra would keep the characteristics of the soloist."

Puckette transferred the project to the Macintosh in order to be able to create the desired graphics interface, while Manoury started composing a piece for piano and the new software, called *Pluton*. Puckette rewrote the software for the Macintosh, while Manoury helped him make the software easy to use and accessible for musicians. The new graphics interface was called Patcher and it revolutionized the way Max® was designed and used. David Zicarelli, a well-known Macintosh programmer, was later added to the creative team and his contribution was critical; he transformed the basic structure of Max® from software that demanded the active participation of a programmer, to software that allowed functionality without one. This was achieved through the visualization of the programming procedure, which lets users create programs by manipulating program elements graphically rather than by command line text. He also created various extensions of Max®, allowing for more intricate use of the software. Max® was published in 1990, instantly becoming a success since it provided easy access to a fairly simple and efficient environment for managing software. Max® made computer music programming easier for musicians who would otherwise not have come in contact with it.

18.3 Max® Environments

Max® consists of five elements: *Max*, *MSP*, *Jitter*, *Gen*, and *Max for Live*. Each object belongs to one of these elements and can be connected to any other object. The first element, Max, deals with basic objects,[2] MIDI and other external controllers, data viewing and manipulation, and user interface arrangement. Users can make their own control interface with full support for MIDI/MPE[3] and OSC protocols[4] by using the Max user interface objects [13].

The second element, MSP, deals with audio signal processing. By using a designated algorithm, automation, and external hardware, users can make their own synthesizers, samplers, and effect processors as software instruments. MSP allows users to work with digital audio using simple oscillators,[5] apply additive (sums of functions, such as Fourier) and modulation synthesis techniques (such as FM synthesis), manipulate audio data, work with filters, perform basic and advanced dynamic processing, work with MIDI, manage resources for creating compound sound patches, integrate multi-channel audio systems, perform audio analysis techniques, use delay,[6] and apply VST (Virtual Studio Technology) effects and instrument plug-ins.

MSP intercommunications set up a relationship among connected objects used to calculate audio information in realtime, whereas Max intercommunications send more abstract messages either cued by users (a mouse clicked, a computer keyboard pressed, a MIDI note played, etc.) or because the event is scheduled in the program[7] [28].

The third element, Jitter, deals with real-time video, 2D/3D vector graphics, and visual effects. Jitter enables users to experiment with audio-to-video

---

[2] Basic objects in Max include *button*, *toggle*, *print*, *message*, *comment*, etc.
[3] Multidimensional Polyphonic Expression (MPE), a method of utilizing MIDI that allows multidimensional devices to take control of multiple parameters of each note within software compatible with MPE.
[4] Created as a successor to the MIDI control protocol, Open Sound Control (OSC) is a communication protocol amongst sound synthesizers, computers, and other multimedia devices optimized for modern networking technology.
[5] Simple oscillators consist of sine, rectangle, square, and triangle waves.
[6] Audio effects such as flangers, echo effects, harmonizers, and reverbs are manipulated by using a delay concept.
[7] For this purpose, we can use objects such as *metro*, *delay*, *pipe*, *drunk*, etc.

control (and vice versa), build video-processing routing and feedback systems by creating and working with matrices, operate live video input and video output components, facilitate MIDI control of video, handle MSP audio in Jitter matrices, draw 3D texts, map spatial settings, and so forth.

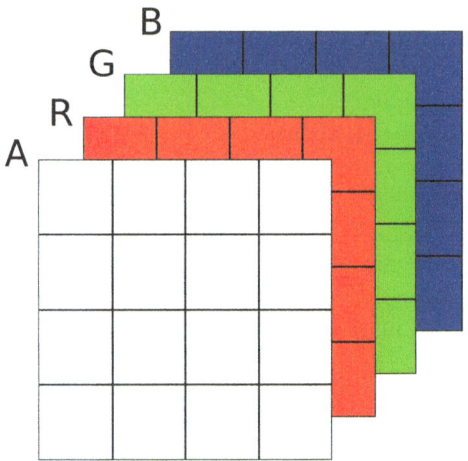

Fig. 18.2: Four virtual planes representing channels for storing numerical values of ARGB.

Jitter generates images by layering matrices (rectangular grids) called virtual *planes* for storing data [29]. This data is stored and arranged in pixels; each pixel is treated as a cell. Each cell has substances of four different values; alpha (A), red (R), green (G), and blue (B), scaled from 0 to 255. These values are treated as channels, assigned numbers 0 to 3, in order to store numerical values of ARGB (0 for A and 1 to 3 for RGB). Breaking down a color into three components (red, green, and blue channels) plus opacity/transparency (the alpha channel) is a standard procedure in a computer to express colors [29]. Therefore the combination of four numerical values stored on four interleaving virtual planes exhibits colors in a two-dimensional video frame (see Figure 18.2).

The fourth element, Gen, is an extension for building highly efficient code from visual graphs and textual expressions. However, Gen enables users to design their own objects without using language-oriented programming. In other words, Gen is meant for people who have hit the limits of Max® as conventional prefabricated software [14].

The fifth element, Max for Live, integrates Max® with Ableton Live® software. It provides access to hundreds of custom plug-ins (known as *Live Devices*) as well as customized tools such as MIDI and audio effects, audio and video synthesizers, 3D Jitter visuals, and tools that interact with Ableton Live® [1, 30].

18.3 Max® Environments

Max® contains high-level objects in addition to objects included in the elements of Max, MSP, and Jitter. *BEAP* (Berklee Electro Acoustic Pedagogy), one of the additional packages in Max 7®, supplies a library of high-level modules inspired by the analog synth environment. The development of BEAP was supervised by Matthew Davidson, a Cycling'74 employee, while he was an instructor at the Berklee School of Music [1], see Figure 18.3. With BEAP modules, users can manipulate sound in multiple ways [13]. Users can drag any module instantly from the BEAP library, drop it into a patch, and connect it with other objects to build a modular performance patch [31]. Because of this kind of operation, BEAP acts as a synthesizer with modules that are integrated with Max.

Fig. 18.3: A BEAP library (from the left toolbar) and the *Phase Vocoder*, one of objects contained in the library.

The other additional package for Max 7®, *VIZZIE*, is a library of high-level image-processing modules. It permits users to create interactive video programs and provides real-time control of little patches [13]. It works similarly to BEAP in that users can drag and drop modules from the VIZZIE library into a patch to manipulate visuals, see Figure 18.4. VIZZIE enables users to work with BEAP modules, make a recording of the VIZZIE patch output, load and play-back movies, manipulate video data, work with MIDI as an input, mix movies, alter video brightness/contrast/saturation, apply video effects, and so forth.

Fig. 18.4: A VIZZIE library (from the left toolbar) and the *4DATAROUTR*, one of its objects.

Objects from BEAP and VIZZIE libraries are actually composed of patches or subpatches that display desired visual elements as new objects. This new object type is known as a *bpatcher* object. It embeds subpatches within a visible and friendly user interface (see Figure 18.5). The number of inlet and outlet objects included in its subpatches window determines the number of inlets and outlets in a bpatcher object [32].

Mira, the other extension in Max®, is a touch-controller application for the iPad that links it to Max® and mirrors its interface. Mira allows users to control any number of patches automatically from an iPad, specify visible regions in a user's patch by the object *mira.frame*, support the majority of standard Max® user interface objects, perform gestural control using the object *mira.multitouch*, dispatch accelerometer data from an iPad using the object *mira.motion*, take control of a single patch for collaborative performance using multiple Mira in multiple devices, and work over Wi-Fi and an ad hoc network [15].

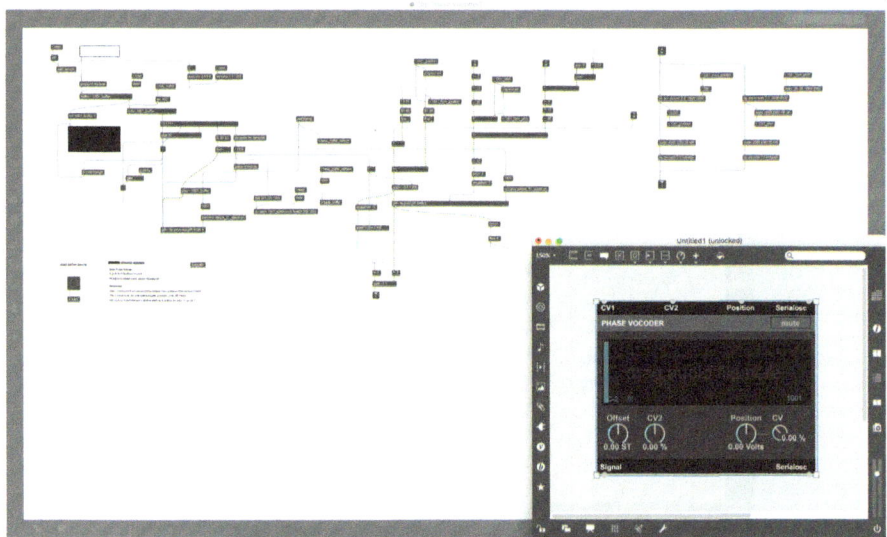

Fig. 18.5: The Phase Vocoder, one of bpatcher's objects (the bottom-right patch) unpacked and viewed in the new way (the larger patch).

18.4 Some Technical Details

To understand the basic idea of the software's usage, let us discuss some important technical details of Max®. This section describes the software's GUI, elaborates the anatomy of an object, analyzes the order of message execution within a patch, and discusses the performance of calculations.

As explained in the previous section, users need a document called a patch where they can arrange objects. A patch displays a rectangular workspace and toolbars (strips of icons) at the rectangle's boundary (see Figure 18.6). These toolbars allow users to work with objects and modify the patch.

An object has the shape of a small rectangle, consisting of an object name, followed by arguments and/or attributes, an inlet, and an outlet. Arguments and attributes function to specify how an object works. First, arguments are set and next, attributes are defined.

Figure 18.7 shows a *cycle~* object, an oscillator for generating a periodic waveform.[8] This object has three choices of arguments, specifying the frequency number, the buffer-name symbol, and the sample-offset integer. It also has five options of attributes specifying the buffer name, the first sample of buffer, the overridden size of the buffer, the oscillator frequency (a floating-point number), and the phase offset.

In Figure 18.8, we can see the already specified object, the metro object with the argument "100" and the attribute "@active 1." The metro object gener-

---

[8] The default waveform of cycle~ is set to a cosine wave.

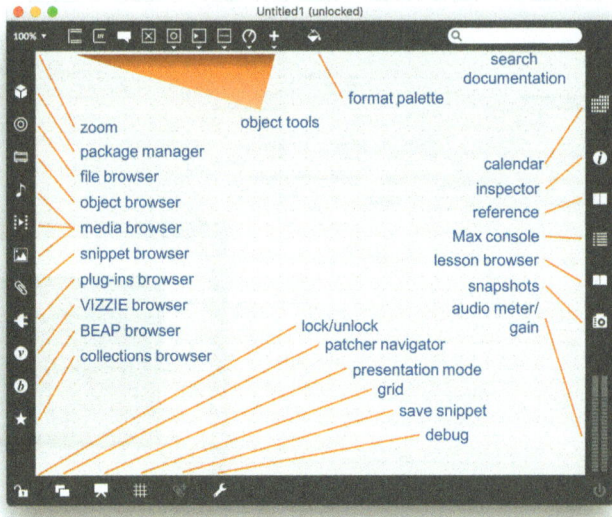

Fig. 18.6: A patch window providing a rectangular workspace and toolbars.

ates a bang message at regular intervals like a metronome; the argument "100" means the interval of 100 milliseconds; the attribute "@active 1" means turning the metro object on.

An inlet receives messages from prior objects and/or outside the patch if an inlet is built in a subpatch (a patch within a patch). An outlet sends messages out to another object and/or outside the patch if an outlet is built inside a subpatch. Such messages carried from outlets are delivered to inlets. Most objects have more than one inlet or outlet, but some objects have only one inlet and one outlet, such as a button or a toggle object. A comment object even has only one inlet and no outlet.

The leftmost inlet of an object is called a *hot inlet*; a message or operation delivered from the prior object to such an inlet is summed to the message in the corresponding object, thereby updating the inlet's corresponding value [10, p. 132]. The other inlets are called *cold inlets*; they store a message from the prior object to the internal corresponding value in arguments and/or attributes so that the message is ready to be applied the next time the message coming to the hot inlet is triggered [10, p. 132].

In Max®, there is an order of message execution, see Figure 18.9. Messages generated in a patch seem to occur at the same time, but in reality they are generated in sequential order. We usually suppose that such actions are ordered from left to right and from top to bottom; but in Max®, the order of messages is executed from right to left and from bottom to top. When a bunch of messages

18.4 Some Technical Details   153

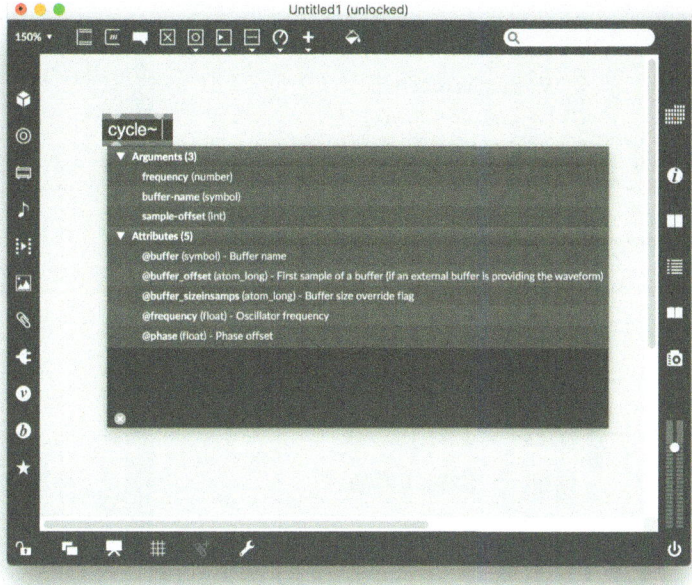

Fig. 18.7: A *cycle~* object with its options to specify arguments and/or attributes.

Fig. 18.8: The metro object with the argument "100" and the attribute "@active 1".

is executed at the same moment, the rightmost/bottommost message will be generated first and the leftmost/topmost message will be generated last [10, p. 109].

Max® contains objects that allow users to perform calculations; these objects include "+" (to add two numbers and output the result), "−" (to subtract one number from another and output the result), "∗" (to multiply two numbers and output the result), "/" (to divide one number by another and output the result), "%" (to divide one number by another and output the remainder), etc., see Figure 18.10. The value of integer numbers is stored in the *number* object whereas the value of floating-point numbers is stored in the *flonum* object. An integer or a floating-point number can also be stored in the *message* object. Objects performing calculations are salient for specifying particular values in determining a task such as defining a sinusoidal-wave frequency, indicating a time occurrence, specifying a precise location, and so forth.

154   18 Max®

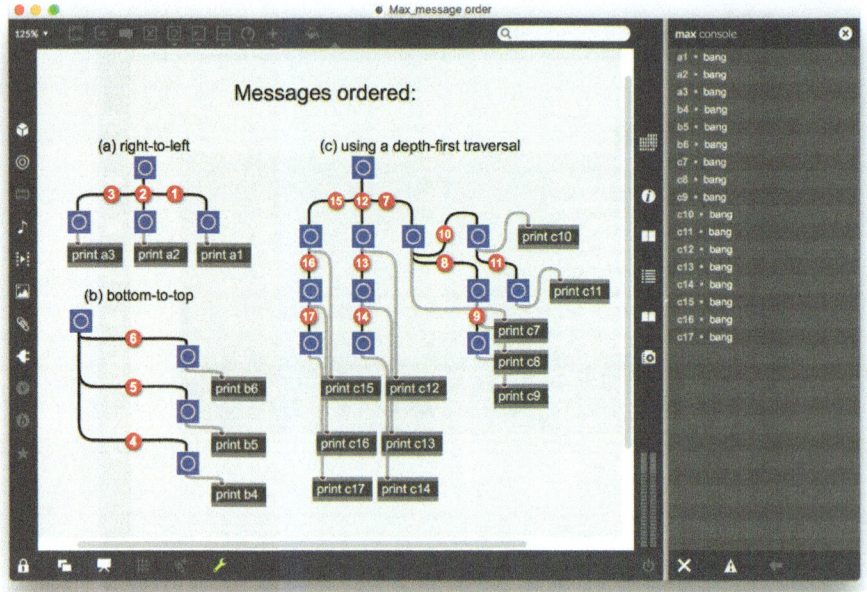

Fig. 18.9: An example of message execution order in Max®. The red-circled numbers show the order of execution.

18.5 Max®Artists

Max® is used in numerous artistic environments to serve a multitude of tasks. There are many sound and visual artists who are famous for their work with Max®. Here we present three of them:

- **China Blue** is known for her work *MindDraw: Theta for Pauline Oliveros* in which theta brainwaves are sonified, and for her work with NASA, which included research into the acoustics of simulated asteroid impacts, see Figure 18.11. Her work is mainly characterized by merging art and science.
- **Mari Kimura** is famous for her work with sensors, see Figure 18.12. She tries to track down human motion and incorporate it in her synthesis. She was introduced to Max® by David Zicarelli at IRCAM. Her newest album *Voyage Apollonian* was made in collaboration with Ken Perlin, a computer graphics artist and technical Oscar winner. He created a fractal transform upon which Kimura composed music. In order to do that, Kimura used motion sensors in combination with Max®.
- **Timothy Weaver** works with installations and live cinema performances, see Figure 18.13. Weaver is a biomedia artist interweaving his microbial ecological and bioenviromental engineering knowledge with sound and visuals.

18.5 Max® Artists   155

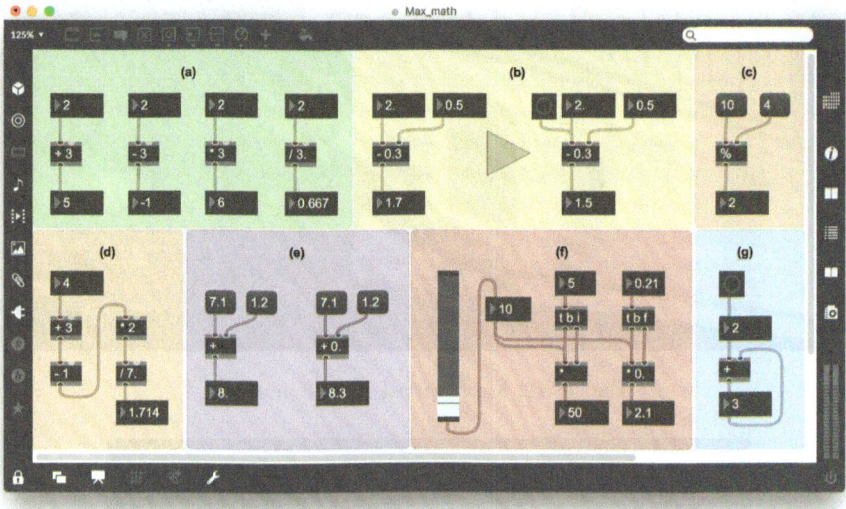

Fig. 18.10: Some patches that perform different calculations: (a) simple arithmetic, (b) hot and cold inlet, (c) modulo operation, (d) order of operations, (e) integer versus floating-point number specified in argument, (f) patch that makes cold inlets hot, and (g) a recursive calculation.

Fig. 18.11: Max® artist China Blue.

Max® plays a central role in his work; it is the platform on which scientific data is transformed into sound and image. Intricate environments are formed as a part of the installation work for the audience's spatio-temporal experience.

Fig. 18.12: Max® artist Mari Kimura.

Fig. 18.13: Max® artist Timothy Weaver.

Part VI

Global Music

19
Manifolds in Time and Space

Summary. When we say "global music," we mean several things. Although "global" refers to things which relate to the world geographically, we can also use "global" to describe things that relate to the entirety of something in general. We explain the concept of global-ness using the Earth as an analogy. When looking at a small local region of the earth, say, your backyard, this region appears perfectly flat. The Earth is a collection of many such regions, but has a new characteristic: it is round. We can therefore say that the earth is a global collection of many flat regions. In each chapter of Part VI, we describe a different set of musical 'regions' and how they contribute to a global music.

In this chapter, we examine how musical works are global because they may be understood both as a whole and as an interaction of many parts. We also discuss musical time in a similar fashion.

$$- \Sigma -$$

19.1 Time Hierarchies in Chopin's Op. 29

In general, we can describe a musical work in terms of space and time. Musical performances often feature spatio-temporal patterns. When we analyze each pattern, we notice that each is simply an arrangement of movements in time. We call these patterns *musical gestures*. Just as many flat surfaces compose a sphere, many gestures construct music. Individual gestures may have local senses of time or motion different from those of the performance as a whole. In fact, when we make music we adjust these movements or the sense of time to the whole. We now talk specifically about global and local time in music.

As we know from reading basic Western music notation, noteheads and flags describe note durations using fractions relative to a whole note. Although this notation can be very precise, it does not describe the literal duration of any note. For example, within Frederic Chopin's *Impromptu* op. 29, there are several different 'local' tempi, and note lengths take on more fluid definitions

160   19 Manifolds in Time and Space

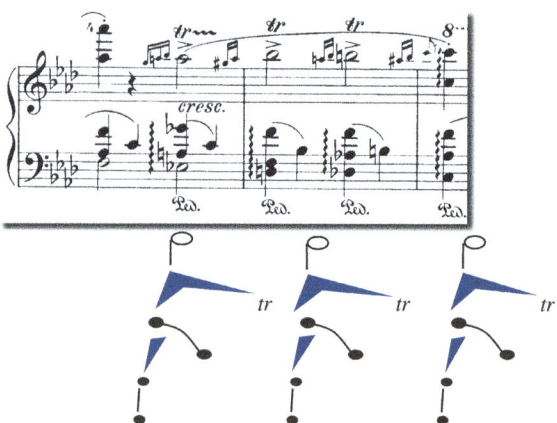

Fig. 19.1: A simple hierarchy breakdown within Chopin's *Impromptu for solo piano* op. 29.

(Figure 19.1). We describe these variations in tempo using the concept of a *time hierarchy*. Time hierarchies are generated by dividing "mother" times (such as the tempo set by a conductor) into "daughter" times that represent local variations. For example, the grace notes may move differently in time than the half note (on beat 3 of the first measure shown.) These two groups therefore form separate daughter times of the mother tempo. Furthermore, two quarter notes are played within the space-time of the half note, creating a granddaughter time that lives within the half. Music consists of not only one time, but a whole genealogical tree of many local time levels. In other words, time is a global construct. The control of such local time hierarchies can be executed by Mazzola's Presto composition software [37, Section 38.2], sound example chopin.mp3 shows four variants of this piece of music with different local tempi.

19.2 Braxton's Cosmic Compositions

> I know I'm an African-American, and I know I play the saxophone, but I'm not a jazz musician. I'm not a classical musician, either. My music is like my life: It's in between these areas. —Anthony Braxton

Anthony Braxton is a multi-talented musician who plays clarinet, saxophone, flute, and piano. Much of his music is based on free jazz-style improvisation inspired by images. The scores for his compositions often resemble colorful paintings. For example, his *Composition 76*, written in 1977 for three "multi-instrumentalists," features a visual title on the cover of the score (see Figure 19.2). The score itself is printed in full color and comes with original liner notes, composition notes, and a notation legend [6].

19.2 Braxton's Cosmic Compositions

Fig. 19.2: Cover page and part of the score of Anthony Braxton's *Composition 76*.

One of his larger ensembles, the Tri-Centric Orchestra, gathered to record his opera *Trillium E* in 2010. Braxton has also written music for multiple orchestras, multiple planets, and even for galaxies! The 2014 *Falling River Music* project included many visual scores, which were played by various small ensembles under Braxton's arrangement.

Braxton's compositions often utilize multiple simultaneous senses of time. This is especially true of his work *Composition 82, For Four Orchestras* (Figure 19.3). According to Braxton, this work is the first completed work in a series of ten compositions that will involve the use of multiple-orchestralism and the dynamics of spatial activity. This work is scored for 160 musicians and has been designed to utilize both individual and collective sound-direction in live performance.

Within *Composition 82*, Braxton indicates that time may be regulated in several different ways. Each section of the music has been designed to emphasize a particular type. The four types of time are:

1. Normal metric time, i.e., all orchestras functioning from the same pulse or time coordinate.
2. Multiple metric time, i.e., two or more orchestras in different tempi.
3. Elastic time, i.e., a time coordinate that is always accelerating or slowing.
4. Rubato time, i.e., a coordinate that allows the conductor to 'draw out' a given section of activity.

Braxton also mixes these time principles. The problem that this 'addition' poses for conducting is solved by the use of television monitors—connecting the conductors—so that sectional 'adjustments' can be regularly dealt with. In this way, the time within *Composition 82* is a global collection of several local times.

Fig. 19.3: Braxton's Composition 82 for four orchestras is meant to be played by 140 musicians.

Additionally, *Composition 82* utilizes a global concept of space [7], giving the piece a 'multidimensional' sound. Each orchestra is positioned near the corners of the performance space while the audience is seated around and in between each orchestra. This arrangement allows each listener to hear a different spatial perspective of the music. In this way, the performance creates a global space out of many individual observers. The placement of activity in this project has been designed to totally utilize the spatial dynamics of *quadraphonic technology* (four loudspeakers). Each orchestra will be heard coming from a separate speaker, and the mixture of events in a given section should give the sense of sound movement through space. This should also be apparent (though to a somewhat lesser extent) with stereo record players as well.

We elaborate on Braxton's works because they utilize global space and time quite deliberately. This does not mean that these ideas are restricted to his compositions. On the contrary, nearly all music has local variations in time or space. We can therefore gain greater understanding of any music by applying these ideas. Global performance space-time is systematically discussed in [37, Chapter 38].

20
Music Transportation

Summary. Prior to the invention of the MP3 compressed audio format, the vast majority of music was either enjoyed live or distributed in various physical formats (described in detail in Chapters 7 and 8). As the Internet gained popularity throughout the 1990s, new digital formats and networking capabilities caused earthshaking changes in music distribution. Technology such as peer-to-peer networking and content streaming now enable us to share, perform, and enjoy music from all kinds of sources anywhere in the world. In fact, it has never been easier or more common to *transport* our music wherever we go. In this chapter, we explore how technology has made music global in the physical sense.

$$- \Sigma -$$

20.1 Peer-to-Peer Networking

Seven years after the release of MP3, personal computing was on the rise. In addition to increases in personal hard disk size[1] and processing power, Internet bandwidth was increasing, allowing data to be exchanged at faster rates. An increasing percentage of people were becoming Internet users, purchasing digital music files, among other types of software and hardware. Compared to physical music formats, digital music was relatively advantageous since it took up less physical space and was easier to carry around. Small, portable MP3 players especially enhanced mobility. Innovations in computer networking also made digital music easy to send or download online for anyone with an Internet connection. The ability to transport music without a physical medium was the primary force in the shift away from CDs, cassette tapes, and other physical formats during the 2010s.

---
[1] The amount of information that can be stored on a personal computer. In 2017, it was not uncommon to have a terabyte (10^{12} *Byte*) of disk space, but in the mid-1990s, 500MB was extravagant.

One complication to the new surge in digital music was the debate over whether MP3 could even be considered an acceptable format. Critics pointed to the fact that compression often reduced audio quality in noticeable ways. However, most music listeners barely noticed a difference. This reality speaks in part to the technological successes of MP3, but also illustrates the public's general preference for quantity of content over quality of sound.

The huge potential of the Internet as a means of music transportation was first realized in 1999, when college freshman Shawn Fanning released *Napster*. The idea for this software allegedly arose from college students' frustration with inconsistent MP3 download links. Fanning's solution was a system that not only guaranteed content, but also allowed users to share huge volumes of MP3 files with unprecedented ease. Anyone with an Internet connection and the Napster application could locate and download their favorite MP3 files.

Napster's power and appeal were largely the result of its *peer-to-peer* (P2P) networking model [49]. On one hand, traditional networks operate on a *client-server* model in which a large, central server stores and catalogs all relevant information, see also Figure 20.1. Individual users, or clients, can merely request information from the server. In peer-to-peer networks on the other hand, every user acts as both a client and a server. This model allows more flexibility and distributes data transfers more evenly across the Internet. Peer-to-peer networking also naturally increases the availability of content, since users generally contribute their own files to the shared library.

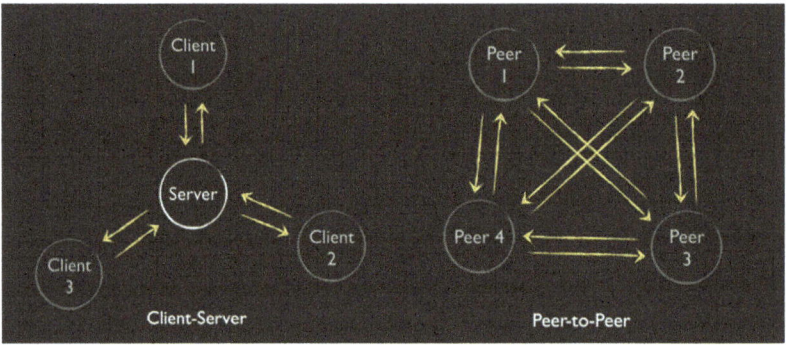

Fig. 20.1: High-level diagram showing the primary difference between client-server networks and peer-to-peer networks. The arrows represent exchanges of information.

Several categories of peer-to-peer networks exist. In *pure* or *unstructured* versions, any user or other entity can be removed without reducing the network's functionality. Links between peers are therefore flexible, allowing users to join and leave freely. However, it can be difficult to find desired content in such a network, since no centralized directory exists. In *structured* peer-to-peer networks, clients keep track of all other clients in the network. This allows quick location of content, but can be computationally intensive and therefore slow. This will be true especially when users are joining and leaving the network at high rates. Finally, *hybrid* networks blend the two, often utilizing a central server to keep track of data while clients store the actual files. The 1999 version of Napster fell into this third category.

By registering with the Napster server, users essentially received access to any MP3 files kept locally on the hard drives of any other Napster users. The condition was that in joining, users agreed to share their own content as well (this arrangement is therefore called *file sharing*). Napster itself did nothing to acquire MP3 files; all content originated from the many users, or peers, in the network. When a user submitted a request for a particular song, the Napster servers sent back a list of peers who owned that song. The user's computer could then assess the various Internet pathways to determine which option would result in the most efficient download time. After selecting a source, the user could proceed to download a copy of the desired MP3 file. The most appealing (and controversial) aspect of the system was that these downloads were free.

From its release, Napster grew rapidly in popularity. This huge transfer of music reached tens of millions of users and resulted in the exchange of hundreds of millions of MP3 files. Users were thrilled. However, many prominent artists, including Metallica and Dr. Dre, angrily found that file sharing cost them significant drops in revenue. These artists and several others sued Napster to have their works removed from the network. Enough former music purchasers had abandoned record shops that these companies saw drops in revenue. Numerous traumatized major record labels sued Napster for copyright infringement. The case ultimately resulted in the website being shut down in 2001. Napster was later bought out and turned into a paid music-streaming service.

But alas, the death of Napster was too late. Almost two years of the appeal and apparent simplicity of free digital music had drastically shifted perceptions of music distribution. The public now had a taste for what technology could offer them in terms of global access to music. Whereas less than a decade before, people were perfectly happy to spend money on their favorite artists' newest albums, listeners could now obtain the same songs for free. Had major record labels acted sooner and perhaps monetized digital music early on, the music industry might look very different today. Instead, music lovers began to question the very concept of paying for music, a sentiment that has yet to be reversed.

20.2 Downloads for Purchase

After Napster, a host of online P2P networks and download sites sprang up throughout the 2000s, many of which were barely legitimate if at all. At this time, piracy became a critical issue for the music industry. As soon as an artist or record label released a song in physical formats, MP3 versions of the files would inevitably end up online, available to be copied by thousands of Internet users. Of course, this was fantastic for some people, especially computer-savvy college students. However, record labels and artists were devastated, experiencing reduced revenue and slacking sales. Much of the technological innovation in music transportation after this time has been intended to recover as much of this lost revenue as possible.

The next step for the music industry was to legitimize the digital distribution of music. The first effective solution came from Steve Jobs, CEO at Apple Computer. After releasing its own line of MP3 players (the *iPod*, see Figure 20.2) in 2001, Apple became the first large tech company to strike licensing deals with major record labels and establish a large library of legitimate music files. Then in April 2003 Apple released *iTunes*, a computer application that provides tools for iPod owners to manage their digital music files. More interestingly, iTunes also provides user-friendly access to the *iTunes Store*, a virtual music shop where users can purchase and download copies of any songs from Apple's library for 99¢ each. Apple's unique strategy was to integrate the iTunes software completely with iPod devices, making it extremely convenient for iPod owners to purchase songs, download them to their personal computer, and transfer them to their device. The combination was extremely popular, with Apple selling its ten millionth iPod and 50 millionth song in 2004 [22]. Apple continued to be a market leader in digital music distribution well into the 2000s.

Fig. 20.2: The first-generation iPod, released in 2001 by Apple Computer.

Convenience and image are arguably the leading factors in the success of the iPod/iTunes combination, and remain essential tools for music distributors and piracy prevention. Convenience is essential because consumers will generally prefer to access music in the way that presents the least cost up front. If legitimate virtual music stores or streaming services are expensive or inconvenient, consumers are likely to prefer piracy. However, if legitimate software such as iTunes provides centralized access and a user-friendly interface, consumers are less likely to pirate and might even be willing to pay. *The iPod's gestural interface creates a unique intuitive consumer experience.* Image is also essential. New companies must advertise to create a cool and popular image, thereby disarming potential pirates. Indeed, Apple's marketing budget greatly exceeded that of its competitors when the iPod was first released.

Upon downloading music files, iTunes users were free to use them however they wanted, with a few limitations. This method of music distribution is commonly referred to as a *product model*, in which music is a good that can be bought and sold in discrete units. This is opposed to the *service model* offered by many streaming companies, in which users pay a subscription to access a large library of music for a limited time period. As with many digital products, consumers wanted the ability to copy songs and play them on all their electronic devices. In order to protect the record labels and artists whose music they were licensing, however, Apple needed to place limits on the number of times a song could be copied and on which devices it could be played. Solutions to this problem are called *Digital Rights Management* (DRM), and usually require both encryption and storage of non-musical data in the song files. Devising suitable DRM technology remains a significant challenge in the music industry.

20.2.1 A Simple Example of Encryption

For readers with little computer science background, we provide a basic explanation of encryption and walk through a very basic algorithm called the XOR cipher. The goal of any encryption algorithm is to deform a piece of information using a predefined procedure and a unique string of numbers or letters, called the *key*. The original piece of information is easily *decrypted* using the key, but recovering the information without the key (i.e., "breaking" or "cracking" the encryption) may be difficult or impossible. Many such techniques exist; the most secure ones typically involve many steps and multiple keys. Apple's DRM uses a relatively complex encryption algorithm, but readers will benefit sufficiently from understanding the XOR cipher as follows:

1. Suppose that the phrase "Go Gophers!" is an important secret that we only want select parties to know. We want to *encrypt* this phrase.
2. We want to use a computer to perform this encryption. This requires us to translate the code phrase into numbers. The ASCII[2] (pronounced "as kee") character set is a widely used way of representing English letters and punctuation as numbers, so we'll represent our secret phrase using these codes. For example, 'G' is ASCII code 71 (01000111 in binary), 'o' is 111 (01101111 in binary) and space is 32 (0100000 in binary). Our phrase in ASCII code is 71 111 32 71 111 112 104 101 114 115 33. A reduced ASCII table is shown in Figure 20.3.
3. We arbitrarily select 'j' as our key for this encryption. The ASCII code for 'j' is 01101010.
4. Several different operations exist that can be used specifically on binary numbers. One such *bitwise* operation is called XOR (read "exclusive or"), in

---

[2] ASCII stands for the American Standard Code for Information Interchange.

Code	Char	Code	Char	Code	Char	Code	Char	Code	Char	Code	Char
32	[space]	48	0	64	@	80	P	96	`	112	p
33	!	49	1	65	A	81	Q	97	a	113	q
34	"	50	2	66	B	82	R	98	b	114	r
35	#	51	3	67	C	83	S	99	c	115	s
36	$	52	4	68	D	84	T	100	d	116	t
37	%	53	5	69	E	85	U	101	e	117	u
38	&	54	6	70	F	86	V	102	f	118	v
39	'	55	7	71	G	87	W	103	g	119	w
40	(	56	8	72	H	88	X	104	h	120	x
41	)	57	9	73	I	89	Y	105	i	121	y
42	*	58	:	74	J	90	Z	106	j	122	z
43	+	59	;	75	K	91	[	107	k	123	{
44	,	60	<	76	L	92	\	108	l	124	|
45	-	61	=	77	M	93	]	109	m	125	}
46	.	62	>	78	N	94	^	110	n	126	~
47	/	63	?	79	O	95	_	111	o	127	[backspace]

Fig. 20.3: A reduced table of ASCII codes showing common characters and the numeric codes associated with them. There is nothing special about the relationship between characters and numbers other than that the majority of computing companies agree to use them.

which the XOR of two identical values is 0, while the XOR of two different values is 1. The table below, called a *truth table*, shows all possible inputs and outputs for XOR.

A	B	A XOR B
0	0	0
0	1	1
1	0	1
1	1	0

We XOR every character in "Go Gophers!" with our key 'j' to get a new set of characters the same length as the original. For example, the first character, 'G' XOR 'j' is encrypted as follows:

$$\begin{array}{r} \text{G } 01000111 \\ \underline{\text{XOR j } 01101010} \\ \text{- } 00101101 \end{array}$$

The resulting ASCII code represents a hyphen. We encrypt the remaining letters in the same way to get a string of character codes that mostly represent non-printable characters (i.e., things like 'backspace' and 'new line'). The ASCII codes in decimal are as follows:
45 5 74 45 5 26 2 15 24 25 75

We now have encrypted our phrase! The encrypted set of codes can now be sent via the Internet to someone else. Other people who know that the key is 'j' can simply XOR each character of the encrypted message with the key again to recover the original. Without the key, decryption is difficult (but certainly not impossible). More complex algorithms are necessary to protect copyrighted music, as in Apple's case.

20.2.2 FairPlay: Fair or Unfair?

Apple's DRM, an encryption procedure called FairPlay, was initially embedded in every file downloaded from the iTunes Store. The music was encoded using the Advanced Audio Coding (AAC) compressed audio format, developed by the MPEG group as an improvement over MP3 (see also Section 8.4). AAC generally maintains better sound quality than MP3, especially at higher compressions. Many tech companies at the time were instead using other algorithms, such as WMA (Windows Media Audio, see also Section 8.4) which also outperforms MP3, but requires its own decoding software. Music distributed by Apple was therefore largely incompatible with other companies' portable music-players and vice versa [23]. On top of using its own encoding algorithm, Apple also developed FairPlay DRM independently of the preexisting Windows Media DRM, invented in 1999, which was available for license at the time. Apple was unwilling to license FairPlay to outside parties, further enforcing incompatibility with other companies' products.

FairPlay encrypted AAC audio using the Advanced Encryption Standard (AES) algorithm. AES results in an encrypted file that can be decrypted using a master key. FairPlay further encrypted this master key with a user key specific to the owner of a given iTunes account. The encrypted audio, the encrypted master key, and information about the consumer were stored in a container file, which is what iTunes users actually received when they downloaded music. Copies of the user key were stored on the iTunes central server, computers authorized under the user's account, and iPods containing the song. The result was music files that could not be decrypted without the user key, which was only available on these devices. Third-party media players could not use FairPlay-protected media. Furthermore, iTunes could not play songs protected by Windows Media DRM.

FairPlay essentially made iTunes music incompatible with everything except iPods and authorized computers. This reality was a source of controversy, as many consumers complained that FairPlay was a restriction while major record labels considered it necessary protection. Regardless of opinions, FairPlay was not unbreakable and many programs to decrypt iTunes files became available over time. Work-arounds were also built into the system, as FairPlay allowed iTunes music to be burned in unencrypted form to a CD. The audio on the CD could then be freely copied by anyone.

Apple first removed DRM from some of its content in 2007, offering this content also at a higher price and audio quality as "iTunes Plus" [23]. DRM was then completely removed from all iTunes music downloads in 2009. iTunes has since added many other types of media including videos, as well as Internet radio and streaming services. Apple continues to be a major player in the field of music transportation.

20.3 The Streaming Model

Even as downloads dominated music transportation, the undercurrent of subscription services began to rise. The idea behind sites such as Pandora, SoundCloud, and Spotify is that users pay a fixed monthly fee in exchange for *access* to a massive library of music (rather than paying for the music itself). This model has become increasingly popular, and represents a shift in the way people perceive and enjoy music. This rising trend also continues to influence consumer expectations. Streaming has also profoundly changed the music industry and has had varying impact on artists.

Whenever data is transported over the Internet it is broken into smaller packets, which then take many different pathways to reach their destination. This concept is fundamental to the Internet and is essential for allowing data to flow smoothly. As a result, when downloading music, the packets are unlikely to arrive in order and cannot be played until all the packets arrive. Advantages of this method are guaranteed seamless playback and offline playback. While the download system was popular through the 2000s and early 2010s, it does present some obvious limitations. For example, the content available on a mobile player is fixed unless connected to a computer.

In streaming systems however, users download and play song files simultaneously. A functional streaming service must solve two important issues: 1) Ensure data packets arrive in the correct order and 2) Ensure data arrives *on time*. Solving both issues requires special networking techniques. The earliest streaming platforms operated similarly to AM/FM radio in the sense that listeners could 'tune in' to a digital stream to hear a live concert or make a conference call. Similar *live-streaming* systems are widely used for television and social media. Live streams are essential when the content is being produced and viewed in real time. However, live streams are less suitable for playing recorded music, especially since they give minimal control to the listener and might break up with poor Internet connections.

Most music services instead use *buffered streaming*. Buffering relies on the fact that musical data can typically be downloaded faster than it is played. Buffered streams download the first few seconds of a song, start playback, then load the next few seconds, and so on. Buffering tends to allow continuous playback, even with relatively poor Internet connections.

Pandora, launched in 2005, utilizes streaming to provide an "internet radio"-type service, where songs appear in sequences defined by genre and user preferences. Users tune in, but have the ability to skip songs, improving control over traditional radio. Pandora was also one of the first streaming services to prominently feature a music recommendation component. New music recommendation has become central to many streaming services, as users now often rely on these services to discover new music. Creating a recommendation engine can be a fairly complex task, and many approaches have been developed. Some services use user data and automate the recommendation process. An algorithm might, for example, simply make advice based on songs that are typically played consecutively. Pandora, on the other hand, implements a more comprehensive approach; professional music analysts rate songs based on a set of 450 attributes, which data are then used to identify similar and dissimilar songs. The service can generate endless playlists of related music, which can be tailored by listeners to achieve a personalized listening experience.

The other popular variant of music service is *on-demand streaming*, in which users can play any songs they want in any order. *Spotify* is an example of a primarily on-demand music service.[3] Spotify uses a unique combination of client-server and peer-to-peer networking to minimize the delay between when users request playback and when playback begins. Spotify's network structure consists of a few central servers and many clients that form a peer-to-peer network. All clients using the Spotify desktop application are part of the peer-to-peer network (mobile users are purely client-server) and store a local cache of music on their personal computer. Cache size is typically limited to 10 GB or 10% of available hard drive space, but the cache is encrypted to protect the music. When a song is requested, the Spotify application first checks the cache for the song. If the song is found, playback can begin immediately. Otherwise, a request is sent to the central server, which sends back the first 15 seconds of the desired song. The client then searches the peer-to-peer network for the remainder of the file. As of 2010, this system results in less than 9% of content going through the central server and has an average latency of about a quarter of a second [24].

20.3.1 Effects on Consumers and Industry

Streaming introduced a new perception of music which appeals greatly to the consumer. This appeal has proved essential in reducing music piracy. With convenient, unlimited access to legitimate libraries of music, the hope is that piracy cannot compete. While piracy certainly has not yet disappeared, copious free versions of streaming services have doubtlessly helped monetize an increasing amount of music enjoyment.

On-demand streaming especially has increased listeners' desires to customize their playlists and listen to wide varieties of music. Before online audio formats, music was typically purchased and listened to in whole albums at a time. Streaming allows and even encourages consumers to pick and choose songs from many albums.

As a result of this apparent pickiness, artists have begun to prefer releasing singles instead of full albums. This tends to benefit smaller artists, for whom streaming is an excellent way to distribute their works to a large audience. However, big-name artists have frequently expressed frustration with streaming services, claiming that the artist payout is simply too small.

---

[3] To clarify, recommendation-based streaming and on-demand streaming are not exclusive. For example, Pandora offers a paid on-demand service and Spotify offers curated playlists.

Overall, streaming is profitable and appealing enough that it will remain a popular means of music transportation. Streaming combines specialized networking capabilities and recommendation engines to provide a modern listening experience.

21

Cultural Music Translation

Summary. There are often clear stylistic differences between works of different cultural origin. Such distinctions are noticeable even between the musics of geographically adjacent countries, such as Germany and France. Put simply, music is full of cultural elements. In this chapter, we consider the music of local cultural regions within music as a 'whole'. This whole comprises all music from every culture. Interestingly, music can be *translated* from one cultural shape to another by use of technology, which emphasizes the global nature of music.

$$- \Sigma -$$

21.1 Mystery Child

The song *Mystery Child* from Beran's and Mazzola's album *Immaculate Concept* [4] is an excellent example of a translation of musical culture. The piece began as Schumann's *Bittendes Kind* from *Kinderszenen*, a suite for piano. Mazzola used the Presto software to deform the pitch and time positions of the original score, then changed the instrument voices. This change works as follows. We first define a regular grid of note events in the plane of onset and pitch. Then we look at every note of the given composition and move it by a certain percentage towards its nearest grid note. This deformation creates a new composition that is a deformation of the original one, but its original structure is not destroyed. Of course, the grid notes will not be heard, they are only a background force. *Mystery Child* is the result; sound example `milesmystery.mp3` presents short excerpts from Miles Davis' *Doo-Bop* CD (see Figure 21.1) and then from *Mystery Child*. It sounds much more like Miles Davis' electric period than a romantic concerto. The parameter manipulations performed in Presto are geometric and can completely change the cultural character of any composition.

Such changes are possible because different cultural musics are connected as a global whole. The geometric operations move the music from one part of the musical globe to another.

174   21 Cultural Music Translation

Fig. 21.1: The cover of Miles Davis' album *Doo-Bop*. Mazzola used the Presto software to transform a Schumann composition into sounds similar to those on this album.

21.2 Mazzola's and Armangil's Transcultural Morphing Software

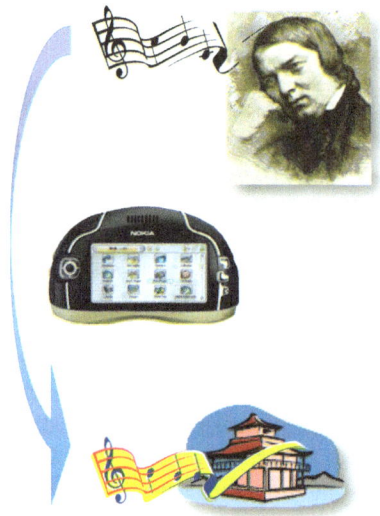

Fig. 21.2: The intercultural morphing project, here from German (Schumann) to Chinese.

21.2 Mazzola's and Armangil's Transcultural Morphing Software

Music translation has potential applications in today's global market. Many international companies seek to reach diverse groups of consumers within each group's cultural context. This can involve tailoring music specifically for each group. Culture is both deep and complex; crossing between any two cultures is challenging. Imagine the possibilities if we could convert music to each listener's appropriate cultural context. In 2003, Mazzola and his student Alev Armangil at the Computer Science Department of the University of Zürich were suggested by Nokia to implement exactly this fascinating desire.

In Mazzola's and Armangil's implementation, deformation algorithms are applied to deform cultural determinants (such as tonalities, rhythms, etc.) to create sounds that appeal to listeners of a desired culture. For example, we can transform a Schumann piece to sound East Asian in tonality and rhythm, but it is still fundamentally based on Schumann; sound example `trafo.mp3` presents the Schumann excerpt followed by the Chinese morphing (see Figure 21.2).

22
New Means of Creation

Summary. In the past, musical creation relied on a special set of performative and theoretical skills. Today, software adds a new dimension to the ways in which music can be created. Software can provide musicians and even non-musicians with easier access to music making. It can offer a friendly workspace, one that requires no lifelong expertise to use. In this way, many more individuals are able to create meaningful music without musical knowledge or instrumental skills. Technology is gradually turning musical creativity into a globally accessible act of self-expression. This new accessibility may open the door to a universal art, where traditional training becomes obsolete and the concept of a specialist outdated. Anyone, anywhere, at any time can use software to become a composer, performer, or artist. However, much as when the invention of cheap cameras brought photography to the public, true artistry may lie beyond software's grasp. For this reason, it is important that we understand the mechanisms by which such technology produces music. In this chapter, we examine a few of the many ways in which music can be produced in a digital environment.

$$-\ \Sigma\ -$$

22.1 The Synthesis Project on the Presto Software

Our first example of global music creation is demonstrated on the jazz CD *Synthesis* [35] which Mazzola recorded in 1990. Its entire structure—in harmony, rhythm, and melody—was constructed using the composition software Presto (see Figure 22.1). Presto was originally written for Atari computers, but now also works on Atari emulators for Mac OS X. The composition was entirely based on 26 classes of three-element motives, i.e., short melodies (see [42, p. 251] for details). The album is a four-part, 45-minute piece; it features piano, bass, and drums, but only the piano part was played by Mazzola. The entire bass and percussion parts were performed with synthesizers, driven by the Presto application via MIDI messages. Through Presto, Mazzola was able to play with a complexity that human percussionists would never be able to interpret from a score.

During the production of this composition, Mazzola never felt inhibited by the predefined electronic soundtrack. On the contrary, he enjoyed 'collaborating' with the complex structures of rhythm and melody. Upon this album's release, it was not recognized as computer-generated music, even by jazz critics; sound example synthesis.mp3 presents a short extract of this music. This album illustrates the power of technology in creating new music. Most notably, this enterprise was a global extension in that it connected human musicianship to electronic music soft- and hardware.

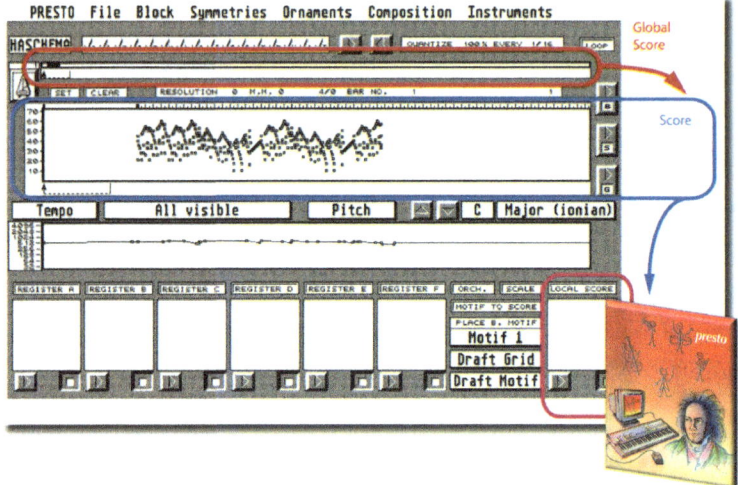

Fig. 22.1: The Presto user interface.

22.2 Wolfram's Cellular Automata Music

In the 1980s, mathematician Stephen Wolfram extensively explored the field of cellular automata, binary computational systems defined by simple rules. He represents these systems as grids of black and white squares, usually starting with a single black square in the center of the top row. Squares in subsequent rows are colored based on the row directly above and a set of 'rules', as in Figure 22.2. By varying the rules, Wolfram found that some of these simple

22.2 Wolfram's Cellular Automata Music

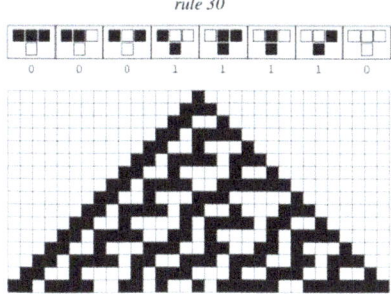

Fig. 22.2: The type of cellular automata studied by Wolfram. Each cell is generated based on three cells in the row above, with the rule indicated in the header.

systems demonstrate mathematically random behavior. This discovery is the basis of his cellular automata for music.

WolframTones, released by Wolfram in 2005, uses cellular automata to create unique cell phone ringtones; see also [58], sound examples `wolfram1.mp3`, `wolfram2.mp3`. The website searches the 'computational universe' for automata with reasonably complex behavior, then renders the patterns as music. Since a near infinity of unique ringtones can be produced, it is unlikely any two users will ever generate the exact same sounds. Wolfram argues that these patterns model creativity, arising from suitably defined automata systems rather than from our human evolutionary history. Furthermore, for Wolfram, the internal consistencies of these patterns give them the potential to render genuinely compelling music. These statements resonate from a mathematical perspective, but many musicians disagree, ultimately finding the musical automata clumsy and monotonous.

An interesting feature of WolframTones is the option to alter both the automata parameters and the constraints used to render them into music (Figure 22.3). Users, regardless of musical training, have the ability to customize their ringtone to their own liking. The website can therefore function as an expressive tool for musicians and non-musicians alike. We call the automata a 'global' means of music creation for this reason.

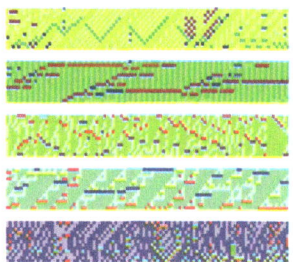

Fig. 22.3: Wolfram's cellular automata 'scores.'

It should be noted that the musical value of a ringtone is determined by the user alone. The algorithm can assess mathematical complexity, but has no way to determine meaning. As a result, the authenticity of the algorithm's creativity remains a topic of debate in the musical and mathematical communities.

22.3 Machover's Brain Opera

The composer Tod Machover believes that anyone can make music. He seeks to explore the mysterious interactions between music, language, and emotion. In 1996 Machover launched a huge project called the *Brain Opera*, with over fifty artists and scientists from the MIT Media Lab. The project incorporated the public into musical creation through intuitive, interconnected gestural applications in a musical laboratory called the *Mind Forest*. It featured various

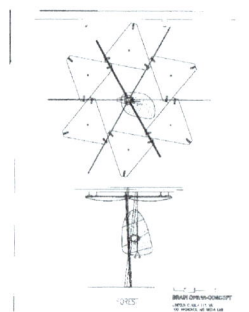

Fig. 22.4: A sketch of a tree from the Mind Forest. Tod Machover, MIT Media Lab.

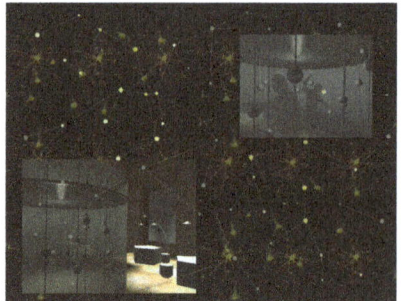

Fig. 22.5: An early artistic rendition of a Rhythm Tree. Tod Machover, MIT Media Lab.

physical devices including fifteen "Trees" (Figures 22.4, 22.5), Harmonic Driving units, Gesture Walls, and Melody Easels. Integrated sensors allowed the public to interact fluidly with the devices. The overall environment was designed to feel natural and responsive so users could express themselves freely. This embedded technology transformed users' gestures and vocal intonations into music and visuals.

People could either sing or speak into these trees. Each Tree would analyze and transform the recording, then store the result for later use in the *Brain Opera Performance*. There was also a "Rhythm Tree", which connected more than 300 drum pads. These pads were made sensitive to variations in touch to add expressive potential. Because they were all connected, each was affected by its neighbors within the web, thus providing even more variation.

Using *Harmonic Driving* units (Figure 22.6), people could direct their way through a piece of music. As if in an arcade game, they would sit in a cubicle with a spring-mounted steering column before them. As described in the project notes, "the player's micro-steering, whether rhythmic and precise or sinuous and meandering, makes the music become sharp-edged or atmospheric." *Gesture Walls* were created to allow full-body motion to control and perturb the music and imagery (Figures 22.7, 22.8). The player's feet would send a small electric signal through the floor plate, which triggered the rest of the

22.3 Machover's Brain Opera   181

Fig. 22.6: An early artistic rendition of a Harmonic Driving unit. Tod Machover, MIT Media Lab.

sensors into activity. Again, these devices stored user input for later use in the performance.

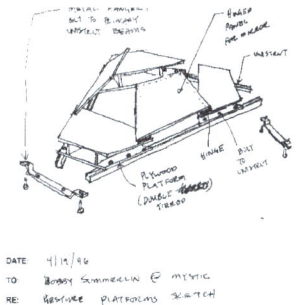 

Fig. 22.7: An early sketch of a gesture wall. Tod Machover, MIT Media Lab.

Fig. 22.8: An early artistic rendition of a Gesture Wall. Tod Machover, MIT Media Lab.

The final feature of the Mind Forest was the *Melody Easel*. Much like finger painting, users were able to drag their finger across a surface to create a melodic line. Their movements and touch created articulation and "timbral filigrees" that were represented in the final sounds and visuals. All of these devices were built to capture the creativity of the people in the Forest, thereby globalizing music creation. Whereas in previous examples technology enhanced the ability of individuals to create a polished creative product, the role of technology in this case was to harness individuals' raw creative impulses. These impulses could then be incorporated into a comprehensive final performance.

Once a group had created enough creative content using the Mind Forest, this group became the audience for the Brain Opera. The pieces they created were assembled using the central, Brain Opera net instrument, called the Palette. This was a Java-based online musical instrument that could be played from one's personal computer. In fact, there were times in the performance of the Brain Opera that these "Internet performers" were the primary focus on the stage. On the dates of the Brain Opera performances, the project website channeled all the data to the Lincoln Center, where the resulting MIDI would be played for the live audience (Figure 22.9).

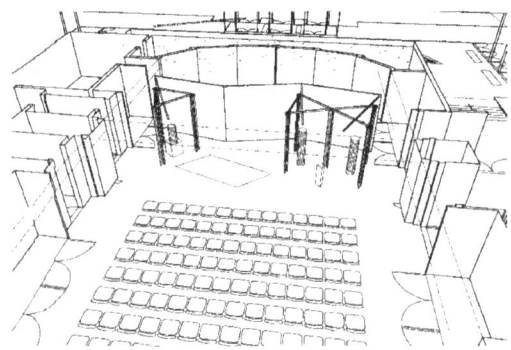

Fig. 22.9: A sketch of the Brain Opera performance space at the Lincoln Center. Tod Machover, MIT Media Lab.

The final performance was a three-movement, 45-minute composition. During the rendition there were three live performers who chose, transformed, and interpreted both composed music and sound improvised by the public audience. The performers directed their way through all of the prerecorded content using tools called the Gesture Wall, the Sensor Chair, and the Digital Baton. The Baton sensed direction and grip, the Chair converted movement into sound, and the Wall was a slight modification of the Wall from the Mind Forest.

The first movement of the Brain Opera was Machover's rendition of Bach's *Ricercare* mixed with the rhythm of spoken words and the audience's recordings. The second movement was heavily instrumental. At the beginning of the third movement, the "hyper-performers" on stage paused, and the internet performers took over. The piece developed into an alternating improvised harmonic flow. Finally, there was the "hyper-chorus climax" when every element was synthesized together in a final and grand recapitulation. The end product was a chaotic cloud of sound. See [26] for more information.

22.4 The VOCALOID™ Software

22.4.1 Introducing VOCALOID™: History

Developed by Yamaha since 2000, *VOCALOID*™ is a voice synthesizer program. The latest version, VOCALOID4™, has been available to the public since 2014. When Yamaha named this software "VOCALOID™", they had a double meaning in mind. First, the suffix "-oid" means "-like," so the name means "voice-like." Second, VOCALOID™ also combines "vocal" and "android", which indicates its technological nature. Either way, VOCALOID™ is a voice synthesizer software that creates virtual singers from sampled human voices.

Users can simulate a voice by simply entering the melody and lyrics into the software. Users can also adjust different dynamic aspects of the voice for different singing and voice styles. Voice banks include various types of voices. In VOCALOID4™, the virtual singer sings Japanese, Chinese, Korean, English, and Spanish. For more references, see [54].

22.4.2 VOCALOID™ Technologies

VOCALOID™'s singing synthesis technology is a type of concatenate synthesis, i.e., it splices and processes fragments of human voices. Lyrics are formed based on phonetic rules, and users can adjust words to different pitches. As of 2017, over a hundred voice banks have been published on the VOCALOID™ platform by different VOCALOID™ companies. Voices are diverse in characteristics such as language and gender.

Similarly to other digital audio workstations (DAWs), the VOCALOID™ software interface is made up of a track editor, a piano roll, a mixer, an effects screen, and a parameter controller (Figure 22.10). To make a virtual singer sing, the VOCALOID™ software enables users to synthesize both lyrics and melody. The main feature of VOCALOID™ software is that users can enter a syllable or a word on each MIDI event, and modulate sound parameters to produce realistic voices.

Music producers control sound parameters to create expressions and dynamics of human voices for live shows or CD recordings. Similarly, VOCALOID™ users can apply sound parameters to make a virtual singer perform more convincingly. Here we discuss the parameters from the most commonly used version VOCALOID3™:

VEL (Velocity): Velocity controls how quickly the voice sings. This works by adjusting the length of the consonants of each syllable. The higher the velocity value, the shorter the consonant, which makes the lyric more attacking. Velocity has a huge effect on fricatives and affricates. [1]

---

[1] Fricative consonants are formed by impeding the flow of air somewhere in the vocal apparatus so that a friction sound is produced. Affricate consonants are formed by stopping the flow of air somewhere in the vocal apparatus and then releasing the air relatively slowly so that a friction sound is produced.

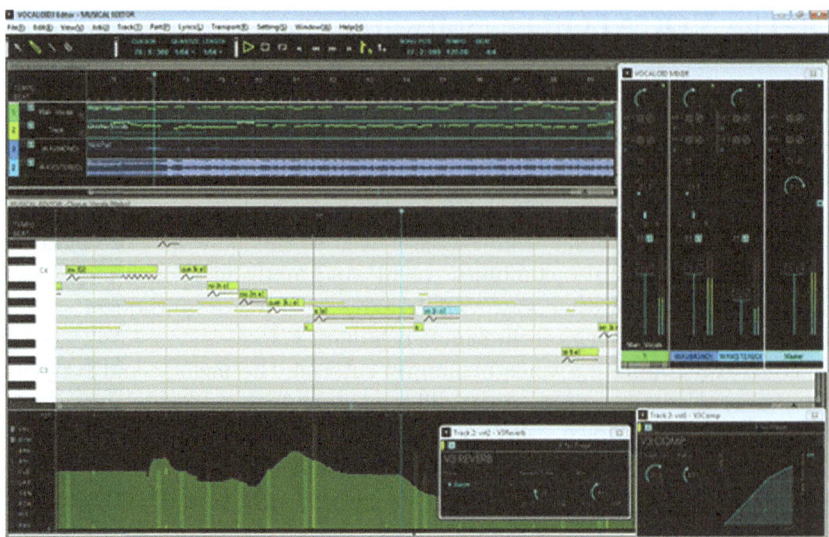

Fig. 22.10: The VOCALOID™ software interface is made up of a track editor, a piano roll, a mixer, an effects screen, and a parameter controller. Users can type words or syllables on MIDI events of the piano roll.

DYN (Dynamics): By adjusting the dynamics curve, the user can tune the singer's volume through the entire track. This is used for adding singing expressions such as crescendo or diminuendo.

BRE (Breathiness): Breathiness regulates the amount of the singer's breath. A higher breathiness adds more breath to the voice.

BRI (Brightness): This parameter changes the voice timbre by boosting or cutting the high frequency components. Higher values produce higher brightness of sound.

CLE (Clearness): This parameter is similar to brightness but interprets voice timbre differently. Clearness determines the sharpness of voice to make it clearer. The higher the value, the sharper and clearer the voice.

OPE (Opening): Opening means the opening of the singer's 'mouth'. This parameter affects pronunciation. A higher value creates a clear and open tone, while a lower value produces a dimmer voice.

GEN (Gender Factor): This parameter adjusts the gender of the original voice. A higher value produces a masculine voice while a lower value produces a feminine voice.

POR (Portamento Timing): Portamento means sliding from one note to another. The portamento timing controls the starting point of the slide.

PIT (Pitch Bend) and **PBS (Pitch Bend Sensitivity)**: These two parameters work together to determine the size of continuous pitch changes. Pitch can be bent by up to ± 2 octaves. These parameters are used to create subtle pitch variations.

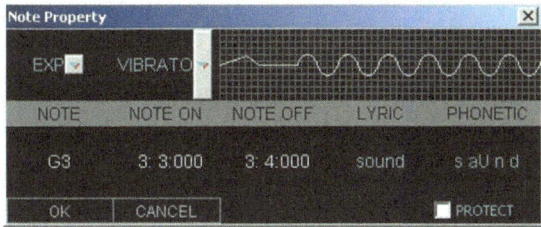

Fig. 22.11: VOCALOID™ note property window.

Besides changing these ten parameters, users can also add vibrato in the note property window (Figure 22.11) to make the singer sound more human. The wave curve shown in the figure indicates the vibrato that is added to a specific note. Users can adjust vibrato length, especially when writing a note with a long duration. A value of 50% means that the vibrato lasts half of the note's duration. Several vibrato types are provided as options; however, users can create their own vibrato styles by changing the vibrato's depth and rate.

All the sound parameters need to be adjusted appropriately since each singer has his/her own voice features and pitch range. Excessive adaptation will break voice originality and decrease authenticity. Skillful use of parameters will produce creative effects such as whispering.

22.5 The iPod and Tanaka's Malleable Mobile Music

In 2004, Atau Tanaka initiated the concept of worldwide social musical collaboration, a vision that is now realistic with the omnipresence of smartphones. Tanaka sought to extend music listening from a passive act to a global activity accessible to the general public. His system exploited specialized handheld mobile devices, as well as ad-hoc wireless networks to allow communities of users to participate in the real-time creation of a single piece of music (Figure 22.12). Some details of Tanaka's plan were as follows:

> The evolution of the music comes from sub-conscious as well as volitional actions of the listeners. The intensity with which a listener holds the mobile device is translated into brightness of the music. The rhythm the user makes as he swings along with the music is captured and drives the tempo through time-stretching techniques. The relative geographies of users in the group drives the mixing of the different musical modules. As a listening partner gets closer, their part is heard more prominently in the mix.

At that time, the technology and social acceptance simply did not exist to make widespread participation possible. Since smartphones were still in their

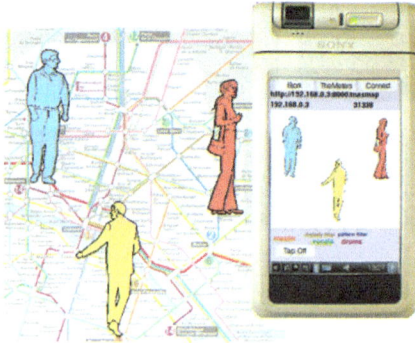

Fig. 22.12: A diagram for Tanaka's Malleable Mobile Music.

infancy, crowd-sourcing[2] and geolocation[3] applications were not as omnipresent or feasible as they are now.

The explosive popularity of Apple's iPod throughout the early 2000s was the first realization of Tanaka's dream. In 2007, the iPod touch was unveiled, which granted users the ability to connect to the internet via Wi-Fi. From there, the global aspect of music, creativity, and communication increased exponentially (see Chapter 20). Many applications, including *Sing! Karaoke* by Smule, and have turned music into a social networking environment. This app prompts users to sing along to popular songs and asks their friends to rate them. The music itself becomes a method of making contact between friends, ultimately making it a method of communication.

---

[2] Crowd-souring generally refers to completing a task through contributions from a group of participants.

[3] Geolocation allows devices to determine their location by communicating with local wireless networks and/or the Global Positioning System (GPS) satellite network.

References

1. Ableton: BEAP - Powerful Modules for Max for Live.
   https://www.ableton.com/en/blog/beap-powerful-modules-max-live/
2. Adorno Th W: Zu einer Theorie der musikalischen Reproduktion. Suhrkamp, Frankfurt/M 1946
3. Agustín-Aquino O A, J Junod, G Mazzola: Computational Counterpoint Worlds. Springer, Heidelberg 2015
4. Beran J and G Mazzola: Immaculate Concept. CD, SToA music 1002.92, Zürich 1992
5. BigBang: https://www.youtube.com/watch?v=ytGcKfhzF2Q
6. Braxton A: Composition No. 76, 1977 http://tricentricfoundation.org
7. Braxton A: For Four Orchestras, Composition No. 82. LP Album. Arista 1978
8. Cage J: ASLSP. http://www.aslsp.org/de
9. Chabade J: Electric Sound: the Past and Promise of Electronic Music. Prentice Hall, Upper Saddle River 1997
10. Cipriani A and M Giri: Electronic Music and Sound Design, Vol. 1. ConTempoNet, Rome 2010
11. Cook P: http://www.cs.princeton.edu/prc/SingingSynth.html
12. Cooley J W and J W Tuckey: An algorithm for the machine calculation of complex Fourier series. Math. Comput. 19(90), 297-301 1965
13. Cycling74: Patch Inside Max. https://cycling74.com/products/max
14. Cycling74: Max 7 Features. https://cycling74.com/products/max-features
15. Cycling74: Mira. https://cycling74.com/products/mira
16. https://www.youtube.com/watch?v=F3rrjQtQe5A
17. Euler L: De harmoniae veris principiis per speculum musicum representatis (1773). In: Opera Omnia, Ser. III, Vol. 1 (Ed. Bernoulli, E et al.). Teubner, Stuttgart 1926
18. Flanagan J L and R M Golden: Phase Vocoder. Bell Labs Technical Journal, Vol. 45(9), pp. 1493-1509, Nov. 1966
19. Fletcher N H and Th D Rossing: The Physics of Musical Instruments. Springer, Berlin 2010
20. Fineberg J: Guide to the Basic Concepts and Techniques of Spectral Music. Contemporary Music Review 19(2), 81-113, 2000
21. Fux J J: Gradus ad Parnassum (1725). Dt. und kommentiert von L. Mitzler, Leipzig 1742; English edition: The Study of Counterpoint. Translated and edited by A Mann. Norton, New York 1971

22. iPod + iTunes Timeline. Apple, 2010
    https://www.apple.com/pr/products/ipodhistory
23. Jozefczyk D: The poison fruit: Has Apple finally sewn the seed of its own destruction? In: Journal on Telecommunications & High Technology Law, 7(2), 369-392, 2009
24. Kreitz G and F Niemelia: Spotify - Large Scale, Low Latency, P2P Music-on-Demand Streaming. KTH - Royal Institute of Technology and Spotify, Stockholm 2010
25. Kronland-Martinet R: The wavelet transform for analysis, synthesis, and processing of speech and music sounds. Computer Music Journal, 12(4), 11-20 1988
26. Machover T: https://operaofthefuture.com/tag/brain-opera/
27. Manzo V J: Max/MSP/Jitter for Music: A Practical Guide to Developing Interactive Music Systems for Education and More. Oxford University Press, New York 2011
28. Max Online Documentation: How MSP Works: Max Patches and the MSP Signal Network. https://docs.cycling74.com/max7/tutorials/03_msphowmspworks
29. Max Online Documentation: What is a Matrix?
    https://docs.cycling74.com/max7/tutorials/jitterchapter00a_whatisamatrix
30. Max Online Documentation: Max For Live
    https://docs.cycling74.com/max7/vignettes/max_for_live_topic
31. Max Online Documentation: What's New in Max 7?
    https://docs.cycling74.com/max7/vignettes/docnew
32. Max Online Documentation: bpatcher. https://docs.cycling74.com/max7/tutorials/interfacechapter01
33. Mazzola G: Geometrie der Töne. Birkhäuser, Basel 1990
34. Mazzola G and E Hunziker: Ansichten eines Hirns. Birkhäuser, Basel 1990
35. Mazzola G: *Synthesis*. SToA 1001.90, Zürich 1990
36. Mazzola G and O Zahorka: The RUBATO Performance Workstation on NeXTSTEP. In: ICMA (ed.): Proceedings of the ICMC 94, San Francisco 1994
37. Mazzola G: The Topos of Music. Birkhäuser, Basel 2002
38. Mazzola G, G Milmeister, J Weissmann: Comprehensive Mathematics for Computer Scientists, Vols. I, II. Springer, Heidelberg et al. 2004
39. Mazzola G and M Andreatta: From a Categorical Point of View: K-nets as Limit Denotators. Perspectives of New Music, 44(2), 88-113 2006
40. Mazzola G: Categorical Gestures, the Diamond Conjecture, Lewin's Question, and the Hammerklavier Sonata. Journal of Mathematics and Music 3(1), 31-58 2009
41. Mazzola G: Musical Performance—A Comprehensive Approach: Theory, Analytical Tools, and Case Studies. Springer Series Computational Music Science, Heidelberg 2010
42. Mazzola G, M Mannone, and Y Clark: Cool Math for Hot Music. Springer Series Computational Music Science, Heidelberg, 2016
43. Mazzola G et al.: All About Music. Springer Series Computational Music Science, Heidelberg, 2016
44. Mazzola G et al.: The Topos of Music, 2nd edition. Springer, Heidelberg 2017
45. Milmeister G: The Rubato Composer Music Software: Component-Based Implementation of a Functorial Concept Architecture. Springer Series Computational Music Science, Heidelberg 2009
46. http://forumnet.ircam.fr/701.html

47. http://www.math.tu-dresden.de/mutabor/
48. Oechslin M S et al.: Hippocampal volume predicts fluid intelligence in musically trained people. Hippocampus, 23(7), 552-558, 2013
49. https://link.springer.com/article/10.1007/s12083-007-0003-1
50. Tan S-L, P Pfordresher and R Harré: Psychology of Music: From Sound to Significance. Psychology Press, New York 2010
51. Thalmann F and G Mazzola: The BigBang Rubette: Gestural music composition with Rubato Composer. In: Proceedings of the International Computer Music Conference, International Computer Music Association, Belfast 2008
52. Thalmann F and G Mazzola: Gestural shaping and transformation in a universal space of structure and sound. In: Proceedings of the International Computer Music Conference, International Computer Music Association, New York City 2010
53. Thalmann F: Gestural Composition with Arbitrary Musical Objects and Dynamic Transformation Networks. Ph.D. thesis, University of Minnesota, 2014
54. https://soundcloud.com/vocaloid_yamaha
55. Wiggins G, E Miranda, A Smaill, and M Harris: A Framework for the Evaluation of Music Representation Systems. Computer Music Journal 17(3), 31-42 1993
56. Wiil U K (Ed.): Computer Music Modeling and Retrieval (CMMR 2003). Springer, Heidelberg 2004
57. Wiil U K (Ed.): Computer Music Modeling and Retrieval (CMMR 2004). Springer, Heidelberg 2005
58. WolframTones: http://tones.wolfram.com

Index

♭, 111
♯, 111
A, 13
c, 19
Ct (Cent), 53
dB (decibel), 13
Δ, 76
$d(z,w)$, 57
e^z, 79
f, 13, 20
$f * g$, 95
$g(t)$, 31
H, 21
Hz (Hertz), 13
k, 19
m, 19
P, 13
Pa (Pascal), 11
Ph, 20
$S(b,a)$, 31
$STRG$, 134
$s(t)$, 31
$T_{3\times 4}$, 57
$W(t)$, 29
$w(t)$, 20
$\mathbb{Z}_3$, 57
$\mathbb{Z}_4$, 57
$\mathbb{Z}_{12}$, 57

A

AAC (Advanced Audio Coding), 85, 169
abstraction, 110
accent, 112
acoustics, 11
ACROE, 32
actions, order of -, 152
ADC (Analog to Digital Conversion), 84
Adorno, Theodor W., 113
ADSR, 22
Advanced Encryption Standard (AES), 169
aerophone, 35
AIFF, 84
air pressure, 11
algorithm
　FFT -, 81
　Yamaha -, 29
Amberol cylinder, 69
amplitude, 13, 20, 22
　quantization, 70
　spectrum, 20
analog, 65
analysis, 23
　Schenker -, 138
analytical hop time, 103
Apple, 166
architecture, concept -, 133
Aristoteles, 133
Armangil, Alev, 175
ASCII, 133, 167
Atari, 177
atomic
　instrument, 20, 23
　sound, 30
attack, 22
Audio Engineering Society, 115
audio file formats, digital -, 84

auditory
    cortex, 11, 16
    masking, 87
    nerves, 11
autocomplementarity function, 60
automaton, cellular -, 178, 179

B

Bach, Johann Sebastian, 182
band pass filter, 96
basilar membrane, 16
basis, orthonormal -, 81
Baud, 118
BEAP (Berklee Electro Acoustic Pedagogy), 149
bell, 36, 40
    tuning, 3
Beran, Jan, 173
Berliner, Emile, 69
BigBang, 143
bitrate, 86, 89
blue noise, 26
brain, 15, 16
Brandenburg, Karlheinz, 85
brass instrument, 39
Braxton, Anthony, 160, 161
breathiness (in VOCALOID™), 184
bridge, 44
brightness (in VOCALOID™), 184
buffered streaming, 170

C

cable, MIDI -, 118
Cadence Jazz Records, 66
Cage, John, 111
    ASLSP, 111
Cahill, Thaddeus, 48
cantus firmus (CF), 59
carrier, 29
CCRMA, 32
CD, 68, 70
cell, hair -, 16
cellular automaton, 178, 179
CEMAMu, 47
chamber pitch, 13
channel, MIDI -, 122
chant, Gregorian -, 110
chest, 44
    voice, 44

China Blue, 154
Chopin, Frederic, 159, 160
Chordis Anima, 32
chordophone, 35
Chowning, John, 27
chronospectrum, 22
circular reference, 138
clarinet, 38
classification of instruments, 35
clearness (in VOCALOID™), 184
client-server system, 164
cochlea, 11, 15, 16
cold inlet, 152
colimit, 132
    form, 136
communication, 5, 81, 115, 186
    MIDI -, 118
communicative dimension of sound, 14
commutative ring, 135
completeness, 133
complex number, 79
complex number representation of sinusoidal functions, 79
compound form, 135
compression, 83
    Huffman -, 83
    Joint Stereo Coding -, 90
    lossless -, 83
    lossy -, 83
    MP3 -, 82
    psychoacoustical -, 85
computational theory, 140
concept
    architecture, 133
    critical -, 27
connotation, 123
consonant interval, 56
construction
    recursive -, 135
    universal -, 135
content, 65
context, 175
controller, 117
convolution, 95
Cook, Perry, 32
Cooley, James W., 81
coordinate, 134
coordinator, 134
cortex, auditory -, 11

Corti organ, 23
counterpoint, 56
    species, 59
CRC (Cyclic Redundancy Check), 89
creation, musical -, 3
creativity, 23
    musical -, 177
critical concept, 27
crowdsourcing, 185
culture, 175
    of music, 173
curve, tempo -, 112
cutoff frequency, 97

D

d'Alembert, Jean le Rond, 132
DAT, 68
database management system (DBS), 131
data flow network, 139
databit, 122
daughter time, 160
Davidson, Matthew, 149
Davis, Miles, 174
DAW (digital audio workstation), 183
decay, 22
deformation, 31, 173
denotator, 131
    name, 133
diaphragm, 44
dichotomy
    Fux -, 59
    strong -, 60
Diderot, Denis, 132
digital, 65
    audio file formats, 84
    encoding of music, 70
digital reverberator, 99
Digital Rights Management (DRM), 167
discantus (D), 59
discourse, 133
dissonant interval, 56
distance, third -, 57
distribution, music -, 163
drum, 41
duration, 13, 22, 112
DX7 (Yamaha), 27
dynamics (in VOCALOID™), 184
Dynamophone, 48

E

ear, 66
    inner -, 15
    middle -, 15
    outer -, 15
Edison, Thomas, 69
Eimert, Herbert, 109
electrical circuit simulation of acoustical configurations, 97
electromagnetic encoding of music, 65
electronic instrument, 45
electrophone, 35
embodiment, 7, 143
emotion, 179
encoding of music, 68
encryption, 167
energy, spectral -, 24
envelope, 21, 27, 30
EQ, 96
equalizer, 97
esthesis, 14
Euler space, 51
Euler, Leonhard, 51, 79
event, sound -, 115
exponential function, 79

F

facts view, 143
FairPlay, 169
Fanning, Shawn, 164
fermata, 112
FFT, 103
    algorithm, 81
FFT (Fast Fourier Transform), 76, 78
fifth, 51
file sharing, 165
filter, 24, 96
    band pass -, 96
    high pass -, 96
    high shelf -, 96
    low pass -, 96
    low shelf -, 96
    notch -, 96
finite Fourier analysis, 75
Flanagan, James L., 101
flute, 36
FM, 14, 27
fold, vocal -, 45
form, 133

colimit -, 136
compound -, 135
limit -, 135
name, 133
powerset -, 136
simple -, 134
type, 134
formant, 45
formula, Fourier -, 79
Fourier, 14
  analysis
    finite -, 75
  formula, 79
  spectrum, 70
  theorem, 19, 23
  transform, 94
Fourier, Jean Baptiste Joseph, 19, 27
fourth interval, 56
frame, 43, 102
  header, 89
  MP3 -, 88
free jazz, 14, 160
frequency, 13, 20, 52
  cutoff -, 97
  modulation, 27
  Nyquist -, 77
  reference -, 52
  sample -, 76
  sampling -, 86
  vanishing -, 94
fret, 40
function
  autocomplementarity -, 60
  exponential -, 79
  non-periodic -, 93
  periodic -, 76
  sinusoidal -, 20
Fux dichotomy, 59
Fux, Johann Joseph, 56

G

Gauss, Carl Friedrich, 81
Gen, 147
gender factor (in VOCALOID™), 184
general MIDI, 123
geolocation, 185
gesture, 109, 113, 115, 159
GIS (Geographic Information Systems), 137

global
  music, 159
  space, 162
  time, 159
glockenspiel, 42
Golden, Roger M., 101
GPS, 185
Gregorian chant, 110
grey noise, 26
grid, 173, 178
group, standardization -, 84
GUI, 141

H

hair cell, 16
hall, music -, 11
hammer, 43
Hamming window, 103
head, 44
  voice, 44
header, frame -, 89
hearing threshold, 87
hexadecimal, 126
hierarchy, time -, 160
high
  pass filter, 96
  shelf filter, 96
hippocampus, 16
Hitchcock, Alfred, 47
horn, 40
Hornbostel-Sachs system, 35
hot inlet, 152
Huffman compression, 83
human voice, 44
hybrid peer-to-peer network, 165

I

ICMC, 32
idiophone, 35
IN port, 118
inlet, 145
  cold -, 152
  hot -, 152
input cable (in Max), 145
instrument, 32
  atomic -, 20, 23
  brass -, 39
  electronic -, 45
  net -, 181

percussion -, 41
reed -, 37
string -, 40
wind -, 36
woodwind -, 39
instruments, classification of -, 35
Internet, 71, 163
internet radio, 170
interval
  consonant -, 56
  dissonant -, 56
iPad, 150
iPod, 166, 185
  mobile device -, 185
IRCAM, 32, 146
ISC (Intensity Stereo Coding), 90
ISO (Int. Organization for Standardization), 84
iTunes, 166
  Plus, 169

J
Java, 141, 181
jazz, free -, 14, 160
Jitter, 147
Jobs, Steve, 166
Johnson, Eldridge R., 69
Joint Stereo Coding compression, 90
just tuning, 54

K
Karajan, Herbert von, 70
key, encryption -, 167
Kimura, Mari, 154
Klumpenhouwer-network, 136

L
language, 179
larynx, 44
layers of RUBATO®, 141
Lewin, David, 136
limit, 132
  form, 135
Lincoln Center, 182
lips, 32
listener, 14
live-streaming system, 170
lobe, temporal -, 16
local time, 159

longitudinal wave, 12
lossless compression, 83
lossy compression, 83
loudness, 13, 134
loudspeaker, 11, 65, 67
low
  pass filter, 96
  shelf filter, 96
LP, 68

M
Mac OS X, 177
Machover, Tod, 179
major third, 51
mallet, 41
Manoury, Philippe, 146
marcato, 112
marimba, 42
masking
  auditory -, 87
  temporal -, 87
mass-spring model, 32
Max®, 145
Max, 147
Max for Live, 147
Max/MSP, 145
Mazzola, Guerino, 4, 113, 131, 160, 174, 175, 177
mechanical encoding of music, 65
membrane, basilar -, 16
membranophone, 35
memory stick, 68
message
  MIDI -, 115, 177
  MIDI channel -, 123
  MIDI Real Time -, 124
  MIDI system -, 123
  MIDI System Common -, 124
  MIDI System Exclusive -, 123
message categories, MIDI -, 123
method, synthesis -, 14
mic, 65
  types
    condenser, 66
    moving coil dynamic, 66
    ribbon, 66
microphone system, 11
middle ear, 15
MIDI, 113, 115, 141

196   Index

cable, 118
channel, 122
channel message, 123
communication, 118
File, Standard -, 125
general -, 123
message, 115, 177
message categories, 123
Real Time message, 124
Specification, 115
Standard - File, 116
System Common message, 124
System Exclusive message, 123
system message, 123
tick, 125
time, 125
velocity, 122
word, 121
Mira, 150
MIT Media Lab, 180
mixer, 65
mobile device iPod, 185
modal synthesis, 32
Modalys, 32
model
  mass-spring -, 32
  product -, 167
  service -, 167
  streaming -, 170
modeling, physical -, 14, 32
modulation, frequency -, 27
modulator, 29
module (functional unit), 145
module (mathematical), 135, 139
mother time, 160
motif, three-element -, 177
mouthpiece, 36
MP3, 72, 84, 163
  compression, 82
  file identifier, 88
  frame, 88
MP4, 84
MPE (Multidimensional Polyphonic Expression), 147
MPEG (Moving Pictures Experts Group), 84
MPEG-1, 85
MPEG-2, 85
MSP, 147

MSSC (Mid/Side Stereo Coding), 90
multiple-orchestralism, 161
music
  digital encoding of -, 70
  distribution, 163
  electromagnetic encoding of -, 65
  encoding of -, 68
  global -, 159
  hall, 11
  mechanical encoding of -, 65
  ontology, 5
  spectral -, 3, 24
  technology, 3
musical
  creation, 3
  creativity, 177
  culture, 173
MUTABOR, 50

N

name
  denotator -, 133
  form -, 133
NAMM (National Association of Music Merchants), 115
Napster, 164
nerves, auditory -, 11
nervous system, 16
net instrument, 181
network, data flow -, 139
neume, 110
neuron, 15, 16
neutral
  level, 14
  niveau, 65
Newton, Isaac, 19
noise, 25
  blue -, 26
  grey -, 26
  pink -, 25
  red -, 26
  violet -, 26
  white -, 25
non-periodic function, 93
nose, 32
notation, 33
  Western -, 109, 113
notch filter, 96
Note ON, 122

number, complex -, 79
Nyquist
  frequency, 77
  theorem, 77

O

object (in Max), 145, 151
object-oriented programming, 133
octave, 13, 31, 51
Oechslin, Mathias, 16
offset, 111
oniontology, 5
onset, 22
onset time, 13
ontology, music -, 5
opening (in VOCALOID™), 184
orchestra, virtual -, 146
order of actions, 152
organ, Corti -, 23
orthogonality relations, 77, 80
orthonormal basis, 81
OSC (Open Sound Control), 147
ossicles, 15
OUT port, 118
outer ear, 15
outlet, 145
output cable (in Max), 145
overblowing, 37
overtone, 20, 27

P

PAC (Perceptual-Audio-Coding), 87
Pandora, 170
parallel fifths, 56
partial, 20
patch, 145, 151
  chord, 145
pedal, 44
peer-to-peer, 164
  network
    hybrid -, 165
    structured -, 164
    unstructured -, 164
percussion instrument, 41
performance theory, 113
period, 13
  sample -, 76
periodic function, 76
phase, 20

spectrum, 20
  vocoder, 100
Philips, 70
phonautograph, 68
phonograph, 69
physical modeling, 14, 32
piano, 43
pink noise, 25
piracy, 171
pitch, 13, 20, 52
  chamber -, 13
  class, 57
  stretching, 100
pitch bend (in VOCALOID™), 184
Pluton, 146
poiesis, 14
poietic, 65, 82
port
  IN -, 118
  OUT -, 118
  THRU -, 118
portamento (in VOCALOID™), 184
postproduction, 67
power object, 132
powerset form, 136
pressure, air -, 11
Presto, 160, 174, 177, 178
principle, time -, 161
process view, 143
product model, 167
programming, object-oriented -, 133
properties in rubettes, 140
psychoacoustical compression, 85
Puckette, Miller, 146
Pythagorean
  school, 3
  tuning, 53

Q

quantization, 84
  amplitude -, 70
  time -, 70
question, open -, 31

R

RAM, 68
rate, sample -, 70
realities, 5
recorder, 36

recursive construction, 135
red noise, 26
reed instrument, 37
reference
  circular -, 138
  frequency, 52
register, 45
relations, orthogonality -, 77, 80
resampling, 101
reservoir technique, 89
reverberation, 98
reverberator, digital -, 99
ring, commutative -, 135
ringtone, 179
RLE (Run Length Encoding), 83
ROM, 68
round window, 16
RUBATO®, 61, 131, 139
rubette, 61
running status, 126

S

sample
  frequency, 76
  period, 76
  rate, 70
sampler, 118
sampling frequency, 86
scalar product, 80
Schenker analysis, 138
Schenker, Heinrich, 138
school, Pythagorean -, 3
Schumann, Robert, 173
score, 109
Seitzer, Dieter, 85
semiotics, 5
semitone, 51
sequencer, 117
service model, 167
simple form, 134
singer synthesis, 32
singing synthesis, 183
sinusoidal function, 20
  complex number representation of -, 79
slave, 116
slide, 40
Smith, Dave, 115
sonification, 144

Sony, 70
sound, 11, 13
  atomic -, 30
  communicative dimension of -, 14
  event, 115
  example, vi
  spectrum, 24
  trumpet -, 22
sound source, 11
SoundCloud, 170
soundboard, 44
source, sound -, 11
space
  Euler -, 51
  global -, 162
species, counterpoint -, 59
Specification, MIDI -, 115
spectral
  energy, 24
  music, 3, 24
spectrogram, 24
spectrum
  amplitude -, 20
  Fourier -, 70
  phase -, 20
  sound -, 24
Spotify, 170
staccato, 112
Standard MIDI File, 116, 125
standardization, 115
  group, 84
standing wave, 36
startbit, 121
status word, 122
statusbit, 122
stopbit, 122
streaming, 170
  buffered -, 170
streaming model, 170
stretching
  pitch -, 100
  time -, 100
string, 40, 43
string instrument, 40
strong dichotomy, 60
structured peer-to-peer network, 164
Studio für elektronische Musik, 109
studio technology, 67
style, Venetian polychoral -, 12

subpatch, 152
substance, 133
support, 22
sustain, 22
Swing, 141
symmetry, torus -, 58
synthesis, 23
　method, 14
　modal -, 32
　singer -, 32
　singing -, 183
synthesizer, 115, 117
synthetic hop time, 104
system
　live-streaming -, 170
　microphone -, 11
　nervous -, 16

T

Tanaka, Atau, 185
tape, 68
technique, reservoir -, 88
technology, studio -, 67
technology
　music -, 3
　quadraphonic -, 162
tempo, 111
　curve, 112
temporal lobe, 16
temporal masking, 87
tenuto, 112
terabyte, 163
theorem
　Fourier -, 19, 23
　Nyquist -, 77
theory
　computational -, 140
　performance -, 113
theremin, 45
Theremin, Leo, 46
third
　major -, 51
　torus, 56
third distance, 57
three-element motif, 177
threshold, hearing -, 87
throat, 32, 44
THRU port, 118
tick, MIDI -, 125

timbre, 24, 102
time, 161
　analytical hop -, 103
　daughter -, 160
　global -, 159
　hierarchy, 160
　local -, 159
　MIDI -, 125
　mother -, 160
　onset -, 13
　principle, 161
　quantization, 70
　stretching, 100
　synthetic hop -, 104
　types of -, 161
topos, 132
torus
　symmetry, 58
　third -, 56
transform, Fourier -, 94
Trautonium, 47
Trautwein, Friedrich, 47
trombone, 40
trumpet, 40
　sound, 22
tuba, 40
Tuckey, John W., 81
tuning, 50, 51, 111
　bell -, 3
　just -, 54
　Pythagorean -, 53
　twelve-tempered -, 51
twelve-tempered tuning, 51
type, form -, 134
types of time, 161

U

U.P.I.C., 47
unity, 132
universal construction, 135
unstructured peer-to-peer network, 164

V

valve, 40
vanishing frequency, 94
velocity (in VOCALOID™), 184
velocity, MIDI -, 122
Venetian polychoral style, 12
vibraphone, 42

vibrato (in VOCALOID™), 185
view in rubettes, 140
violet noise, 26
violin, 32
　family, 41
virtual orchestra, 146
virtuosity, 32
VIZZIE, 149
VL-1 (Yamaha), 32
vocal fold, 45
VOCALOID™, 182
voice
　chest -, 44
　head -, 44
　human -, 44

W

wave, 12, 20
　longitudinal -, 12
　standing -, 36
waveguide, 32
wavelet, 14, 30
Weaver, Timothy, 154
Western notation, 109, 113
white noise, 25
Wille, Rudolf, 50
wind instrument, 36

window
　oval -, 16
　round -, 16
WMA, 169
Wolfram, Stephen, 178
WolframTones, 179
Wood, Chet, 115
woodwind instrument, 39
word
　MIDI -, 121
　status -, 122
work, 14

X

Xenakis, Iannis, 47
XML, 131, 141
XOR, 167
XOR cipher, 167
xylophone, 42

Y

Yamaha, 27, 182

Z

Zarlino, Gioseffo 56
Zhu, Zaiyu, 51
Zicarelli, David, 146

The manufacturer's authorised representative in the EU is Springer Nature Customer Service Centre GmbH, Europaplatz 3, 69115 Heidelberg, Germany. If you have any concerns regarding our products, please contact ProductSafety@springernature.com

Printed and bound by CPI Group (UK) Ltd, Croydon, CR0 4YY

23/03/2026

02076667-0013